DRINKING WATER CHEMISTRY

A Laboratory Manual

DRINKING WATER CHEMISTRY

A Laboratory Manual

Barbara A. Hauser

LEWIS PUBLISHERS

A CRC Press Company

Boca Raton London New York Washington, D.C.

Library of Congress Cataloging-in-Publication Data

Hauser, Barbara A.
Drinking water chemistry : a laboratory manual / Barbara A. Hauser.
p. cm.
Includes bibliographical references and index.
ISBN 1-56670-486-3 (alk. paper)
1. Drinking water—Analysis—Laboratory manuals. 2. Water quality—Measurement—Laboratory manuals. I. Title.

TD380 .H43 2001
628.1'61—dc21 2001029983
CIP

Direct all inquiries to CRC Press LLC, 2000 N.W. Corporate Blvd., Boca Raton, Florida 33431.

Visit the CRC Press Web site at www.crcpress.com

Lewis Publishers is an imprint of CRC Press LLC

International Standard Book Number 1-56670-486-3
Library of Congress Card Number 2001029983

Printed on acid-free paper

The Author

Barbara Hauser has been department head and chief instructor of the Water Technology Department at Bay de Noc Community College, Escanaba, Michigan, for the past 17 years. Her background includes biochemistry research at the Rockefeller Institute in New York City and at the University of Maine in Orono. She has an A.A.S. in water technology from Bay de Noc Community College, a B.S. in biology/chemistry from Trinity College, Washington, D. C., and an M.S. in occupational education from Ferris State University, Big Rapids, Michigan.

A licensed operator in water and wastewater treatment in the states of Wisconsin and Michigan, she has worked for the Rexnord Company, Milwaukee, Wisconsin, as an industrial operations specialist in advanced wastewater treatment, and for the city of Gladstone, Michigan, in municipal wastewater treatment.

Mrs. Hauser has developed and operated onsite training courses for industrial wastewater treatment personnel, taught short courses and correspondence courses for waterworks operators sponsored by the Michigan Department of Public Health, organized regional waterworks and wastewater operator training meetings, and conducted research for the Michigan Department of Natural Resources.

Mrs. Hauser is the author of the *Practical Hydraulics Handbook*, *Hydraulics for Operators*, and the *Practical Manual of Wastewater Chemistry.* In 1990 she was the recipient of the Outstanding Occupational Educator Award from the Michigan Occupational Dean's Council.

Introduction

A colorless, odorless, tasteless liquid, water is the only common substance that occurs naturally on the earth in all three physical states: solid, liquid, and gas. Approximately 73% of the earth's surface, almost 328 million cubic miles, is covered with water. The human body is 70% water by weight, and water is essential to the life of every living thing.

Water is a unique molecule; one side of it is positive, the other is negative. Water molecules orient to neutralize electric charges; this buffering action keeps foreign ions from reacting and precipitating. Pure water itself resists ionizing, and is a poor electrical conductor. These properties give water an amazing capacity to dissolve other molecules. It is, in fact, considered the universal solvent.

This manual was prepared to enhance the water treatment plant chemist's understanding of laboratory theory and procedure, to appreciate the nature of the contaminants under analysis, the analytical methods, and their limitations. It is important to interpret the data and apply the information for treatment process control. Test results are a reliable description of the sample analyzed; they define the status of the treatment process. Treatment can be optimized and process upsets identified. State and federal regulatory compliance will be achieved.

Methods in this guide are approved by the Environmental Protection Agency (EPA). Included are explanations of all analyses and the equipment used to perform them, to help in understanding the chemistry of the reactions and the working mechanism of the instruments. A summary of treatment plant significance for each analysis is provided. At the back of the guide are typical math exercises, arranged by chapter.

The Safe Drinking Water Act

To protect the American public from health hazards associated with contaminated drinking water, Congress passed the Safe Drinking Water Act of 1974. This was our first national legislation whose purpose was to safeguard public drinking water supplies. Under this law, the EPA had the responsibility of establishing regulations defining safe drinking water quality for public water systems, and assuring that all public water systems provide water meeting this definition. Most states have since assumed "primacy" for the enforcement of the law within their jurisdictions, and must provide regulations at least as stringent as those proposed by EPA.

Under the provisions of this law, state and federal regulations were mandated for **Primary Maximum Contaminant Levels,** based upon public health significance. Upper limits were set for total coliform, certain organics, inorganic metals, nitrate, turbidity, and radionuclides.

Secondary Maximum Contaminant Levels, based upon the water's aesthetic value, for the comfort of the traveling public, and for protection and maintenance of the water distribution system, were strongly advised. This category included sulfate, fluoride, odor, color, chloride, total dissolved solids, corrosivity, zinc, manganese, iron, copper, and foaming agents.

Since its enactment the federal law has twice been reauthorized, and some regulated contaminant limits have changed. Total coliform has changed; fluoride is listed as both a primary and secondary MCL; turbidity regulation has been set into a different category; radionuclides are updated; and new contaminants have been added, including nitrite, arsenic, total trihalomethanes, and many more organics. New operator certification requirements have been established. Specialized "rules" are now enforced (surface water treatment, lead and copper, and disinfection byproducts). Requirements for public notification upon noncompliance have become stricter, and a mandatory provision that utilities provide ongoing public education to verify consumer confidence in the quality of the supplied water is now in place. Revolving loans are available for new construction and expansions.

Definitions

Potable Water: Water that is safe from biological pathogens and chemical contaminants, which are hazardous to health.

Maximum Contaminant Level (MCL): The maximum permissible level of a contaminant in water delivered to any user of a public water supply.

Primary Maximum Contaminant Levels: Limits set on bacteriological, chemical, and physical contaminants that are of health significance. These are **Primary Drinking Water Standards**. They are enforceable.

Secondary Maximum Contaminant Levels: Chemical and physical limits set on aesthetic water quality; these limits represent goals and are not enforceable. These are **Secondary Drinking Water Standards**.

Maximum Contaminant Level Goal: A projected limit based on health significance; a goal to be attained, but not yet enforceable. These contaminants are still under study.

Public Water System: Required to sample and analyze for drinking water contaminants on a regular basis based on the type classification and the source of drinking water. They are divided into three categories: community water systems; nontransient, noncommunity water systems; and noncommunity water systems.

Community Water Systems: Serve a residential population of 25 year-round residents or have 15 service connections (includes municipal water utilities, mobile home parks, condominiums, etc.). All community supplies are classed as **Type I** public water supplies.

Nontransient, Noncommunity Water Systems: Serve individuals for at least 6 months of the year (day-care centers, schools, or factories). These are also **Type I** public water supplies.

Noncommunity Water Systems: Serve transient populations (restaurants, motels, campgrounds, or highway rest areas where people only stop briefly, or reside for a short time). All noncommunity supplies are classed as **Type II** public water supplies.

Regulatory limits listed in this guide apply to Type I public water supplies.

Contents

Chapter 1

Sampling

Contents

All laboratory testing depends upon proper sampling technique. Sampling must be designed to obtain accurate data for identifying treatment process changes and varying water qualities. The objective is to remove a small portion that is representative of the entire flow and adequately reflects actual conditions in the water.

Sample Location

The test requirements usually determine sample location. Obtain the sample where mixing is best and the water is of uniform quality. The sampling location must be accessible. Avoid slippery surfaces. Do not climb on or under guardrails.

Composite Samples

This type of sample is taken to determine average conditions in a large volume of water whose chemical properties vary significantly over the course of a day. Small aliquots are taken at regular intervals and pooled into one large sample over a 24-h period. If composite samples are to be taken manually, the frequency of sampling should depend on the number of changes in water quality. Most often, grabs are taken hourly, and the volume of each is flow proportioned. Samples are then combined into a composite for testing.

Grab Samples

A grab sample is taken all at once, at a specific time and place. Insert container upside down into the water. Rotate open end toward direction of flow and allow to fill under the surface. Sample about 12 in. underwater, or at center of the channel, about medium depth from the bottom. Avoid surface scum and bottom sediment. Be careful not to collect deposits from tank sidewalls. Eliminate any large, uncharacteristic particles that enter the sample. Fill sample bottle completely to exclude air space if sample is to be analyzed for DO, pH, chlorine residual, alkalinity, or VOCs. If preservative or dechlorinating agent has been added to empty sample bottle, adjust sampling technique so that the bottle does not overfill, or the chemical will be washed out. Be careful to handle sterile bottles for bacteriological testing aseptically. Collect enough sample to allow duplicate and spiked analysis. If both influent and effluent samples are desired from a process unit with a detention time of 2 h, consider the timing. Take the influent sample now, and the effluent sample in 2 h. There will be a good chance of getting the same water.

The Sample Bottle

For chemical testing, the sample bottle must be clean. For bacteriological testing, the sample bottle must be clean and sterile. Bottles may be glass or plastic for most analyses; labels must be firmly attached to the sample bottle, not to the lid. Use labels that will not come off when damp. Use a water-insoluble ink pen.

The label on a sample bottle should include sample ID number, date and time of collection, type of sample, location, adverse weather conditions, collector's initials, analysis to be performed, and sample preservation, if any.

Preservation and Transport

Dissolved oxygen, pH, and temperature should be analyzed on site, at the sampling location. Bacteriological tests must be performed within 24 h of sampling. All samples should be analyzed as soon as possible after collection. **See Appendix II for sample preservation and holding times**.

Chain of Custody

This is a legal requirement that refers to the recorded handling of a sample from collection to analysis. It allows for tracing individual samples in case of a problem.

The sample is in custody if it is in hand or in sight, if it is locked away, or if it is placed in a secure area nobody can enter without the possessor's knowledge. When the sample changes hands, a change of possession form is signed by both parties; the date and time are recorded.

The chain of custody records include:

- Sample labels — For sample identification
- Sample seals — For shipped samples, to ensure no tampering
- Field logbook — Includes all information on label, container type, sample size, field analysis, number of samples taken
- Chain of custody record — Includes label information and change of possession forms

Online Sampling and Analysis

There is increasing, and for some parameters mandatory, use of automatic online sampling and monitoring. It has the advantage of yielding constant and immediate results. Time is saved, and problems due to carriage to laboratory are eliminated. Meters are installed directly into the process flow, and the water is monitored as it flows by. Typical tests that are performed this way are pH, DO, temperature, turbidity, and chlorine residual. The signal may be read on site, or transmitted to the lab or control room for remote readout. Frequent cleaning and calibration of the monitoring instruments are essential.

Laboratory Records

Process control and regulatory compliance depend upon the proper recording of laboratory analysis data. Analysis reports (bench sheets) should include:

- Name, time, and date of analysis
- Analyst name
- Sample preparation
- Analysis method
- Test conditions (standards, reagents, instrument settings, temperature, and reaction time)
- Results of analysis
- Observations — comments

NOTE: Laboratory records must be kept for 5 years.

Chapter 2

Laboratory Safety

Contents

Safety is just as important in the laboratory as in the treatment plant. A number of hazards exist. Be alert, careful, and aware of potential dangers at all times. Although very dilute solutions are being analyzed, the laboratory chemist does handle some extremely strong acids and bases, as well as some toxins. A thorough knowledge of safe operating procedures for each analysis is vital.

Laboratory Safety Rules

1. **Know where the safety equipment is.** Essential equipment includes safety shower, eye wash, fire extinguisher, telephone, first aid kit, and acid spill kit. Emergency response must be automatic because it may be difficult to think clearly.
2. **No eating, drinking, or smoking in the lab.**
3. **Don't smell or taste any chemical.**
4. **Use pipet bulbs for pipetting.**
5. **Never handle chemicals with your hands.** Use a spatula.
6. **Never stopper a flask with your thumb to mix the contents.** This is unacceptable laboratory technique. It will contaminate the test, and chemical in contact with the skin or mouth could be dangerous.
7. **Wash hands well with soap and water before leaving the lab.** There are chemical toxins and biological pathogens in the lab.
8. **Wear safety glasses in the lab at all times.**

9. **Label all prepared solutions properly.** Include chemical name, concentration, date prepared, and chemist's initials. If a solution is over 1% concentration (0.1% if it is a carcinogen), OSHA requires that labels also list any fire and health hazards. Special labels may be purchased for this purpose. Most liquids in a water laboratory look like water. If the concentration is not on the label, the liquid is useless as a chemical and could be dangerous.
10. **Add acid to water!** When diluting concentrated acids, always put the water in the beaker first, then add the acid, SLOWLY. Acids are hydroscopic; they react quickly with the water they are dissolving in, creating heat. When a large quantity of acid mixes with a little water, the reaction can be violent.
11. **Use strong oxidizing agents carefully.** Agents such as ammonium persulfate can produce violent reactions. Store separately.
12. **Keep highly reactive chemicals out of sinks.** Chemical incompatability resulting from the combination of oil, grease, mercury, volatile solvents, and strong acids may trap vapors in the drains and create an explosion.
13. **Be aware of heat!** Don't touch a hot plate to see if it is on. Assume that it is. Never leave a heating solution unattended. Dissolve strong acids and bases slowly. They create great heat and may spatter or break the glassware. When opening a furnace, oven, or hot water bath, stand a distance away. Always use tongs and wear gloves.
14. **Strong acids and bases emit choking fumes.** Use them under the fume hood. Do not lean over a boiling solution. It may be emitting toxic vapors. Boil solutions under the fume hood. If you unexpectedly encounter fumes, rush to the nearest source of fresh air.
15. **Acids burn!** Strong acids and bases are highly corrosive, especially to the skin. Handle with extreme care to avoid contact. Wash off with plenty of running water. Use the safety shower. If acids or bases make contact with the eyes, use the eye wash. Call for help. Seek immediate medical attention.
16. **Carry all large chemical containers with two hands to minimize risk of dropping.** Hands and/or glassware are often wet.
17. **Don't try to pick up broken glass with bare fingers.** Sweep it up with a broom. Dispose of it in a special container labeled "broken glass."
18. **Some chemicals are toxic!**

    Arsenic: Highly toxic. Avoid inhalation, ingestion, and skin exposure. Prepare in a fume hood.

    Azides: Sodium azide (NaN_3) is toxic and reacts with acid to produce the more toxic hydrazoic acid. Avoid inhalation, ingestion, and skin exposure.

    Cyanides: Most are toxic. Avoid ingestion. Do not acidify cyanide solutions; toxic HCN gas is produced.

    Mercury: Liquid mercury is a highly toxic and volatile element. Avoid inhalation and skin exposure. Mercury compounds are also toxic.

    Organics: Many are toxic and/or carcinogenic. Many are flammable or explosive.
19. **Beware of electrical hazards.** Do not plug in electrical equipment with wet hands. When turning on a hot plate, ensure that no electrical cords are touching the heater plate. The cord may melt and cause an electrical short.
20. **Water may contain pathogens.** Use medical lab gloves when handling samples, especially if you have a hand cut or abrasion, Anyone working in a water treatment plant should update immunizations with the local Health Department.

21. **Compressed gas is not a toy.** Keep cylinders chained to the wall. Do not fool with pressure regulators. Always open cylinder valves slowly.
22. **Don't try to change the lab test procedure.** Chemicals added, subtracted, or mixed in an order different from the order specified may cause explosive conditions.

> **NOTE: Good housekeeping is part of safety. Be scrupulously clean with the glassware and chemicals. Contaminated glassware will ruin the lab test and can be hazardous.**

Chapter 3

Weights, Volumes and Concentrations

Contents

A water laboratory is precisely quantitative; solutions are prepared in concentrations of parts per million. Every milligram counts, and the results of measurements will depend upon the absolute accuracy of the chemical preparations.

Analytical Balances

The analytical balance is a delicate instrument. Its operation is based on automatic internal setting of weights equivalent to the weight put on the pan to achieve a "balance." It is extremely accurate, and enables weights to be prepared to the milligram (0.001 g). The balance must be kept scrupulously clean. Corrosion from spilled chemicals will restrict the free movement of the pan and will work its way into the operating mechanism.

Troubleshooting Tips

1. Calibrate the analytical balance against standard gram weights often.
2. Do not weigh anything that is not room temperature. Cold substances will weigh more. Hot substances will weigh less.
3. Close balance doors when weighing. Drafts, uneven counters, or nearby bumps and bangs will disturb the weighing operation.
4. Begin by zeroing the balance.
5. Analytical balances are constructed for small weighing operations. Do not overload the balance with a weight too near its maximum capacity.
6. Leave the balance clean with the doors closed when finished.
7. If the weight required is too small to weigh accurately on the analytical balance (<10 mg), prepare the solution in a greater concentration than needed and then dilute it.

Desiccator

A desiccator is an airtight container used to keep substances dry while they are cooling. It has a suspended floor, under which is a chemical desiccant that absorbs moisture from the air. Desiccant should be replaced periodically. Indicating desiccants change color when they are spent. Desiccant chemicals include $CaCl_2$, $CaSO_4$, NaCl, and silica gel.

Volumetric Measurements

Volumetric measurements are just as important as gravimetric ones. If 2.5416 g of chemical is weighed out accurately and dissolved into a liter of water — which isn't really a liter because the volume wasn't measured correctly — the concentration of the solution will not be accurate.

Each piece of volumetric glassware is for a specific purpose. Beakers are meant only for mixing chemicals and should not be used for measuring volumes. Beakers have the least calibration accuracy of all laboratory glassware. Erlenmeyer flasks are to be used for titration; the narrowed tops are designed to keep the liquid from splashing out of a swirling flask. These flasks are not accurate enough to use for measuring liquids. Volumetric flasks, graduated cylinders, and pipets are the most accurate, and are meant for measuring liquid volumes. Use these whenever exact volumes are indicated.

NOTE: Don't measure the volume of anything in a beaker!

Pipets

Some pipets are calibrated down to the tip. When using this type, evacuate all the liquid and touch the tip of the pipet to the side of the container. Some calibrations end before the narrowing at the bottom of the pipet; when using these, measure only down to the end of the calibration. A volumetric pipet has an expanded section in the middle. It is the most accurate, but only one volume can be measured with it.

Temperature

Liquid chemical should always be at room temperature when measuring. Water expands a great deal when it is hot, and later will contract, making the volume measurement deficient.

Relative Volume

Choice of glassware size should be based on common sense. To stir 500 ml of liquid, choose a 1-l beaker. When titrating into a 100-ml sample, put the sample into a 250-ml Erlenmeyer flask. To measure liquids of 10 ml or less, use a pipet. Volumetric pipets are the most accurate. When measuring in a graduated cylinder, use the size that is just a little larger than the volume of liquid needed. A buret is an awkward but very accurate means to measure small amounts of liquids. Burets are used for titrating.

Reading the Meniscus

Water molecules are attracted to the molecules of the container they are in, and the water surface will "climb up" the sides a bit forming a curve, or a meniscus. Read the water level at the bottom of the meniscus. This applies to reading pipets, burets, graduated cylinders, and volumetric flasks.

To Prepare an Exact Concentration of a Solution

Weigh the dry chemical on the analytical balance. Dissolve the chemical in a beaker with about half the amount of distilled water required. Pour into a volumetric flask that is calibrated to the right volume. Rinse the beaker out into the flask with a little distilled water (transfer quantitatively). Fill the flask to the calibration line with distilled water. Cap it and mix well! Turn upside down slowly at least 20 times. It is difficult to mix liquids well in a full volumetric flask.

NOTE: Do not pour dry chemical down the neck of a volumetric flask!

Concentration

Concentration indicates how much solute is contained in a given amount of solution.

Parts per Million

This is a weight-to-weight ratio that relates to parts of the chemical or physical constituent, per million parts of water, by weight. Take fluoride, for example.

10 parts per million is:

10 lb F per million lb water
10 tons F per million tons water
10 g F per million g water
10 mg F per million mg water

These are all the same concentration, and they are all weight/weight. Since by weight a liter of water is composed of 1 million mg water, then:

10 mg F per million mg water = 10 mg F per l water

10 ppm = 10 mg/l

Micrograms per Milliliter

The reference book *Standard Methods* frequently uses the less familiar unit label micrograms per milliliter, or μg/ml. This is also a ppm designation, and is exactly the same as mg/l. It is sometimes used for designating small concentrations. To demonstrate how μg/ml and mg/l relate to each other: a microgram is a thousand times smaller than a milligram; if the unit label is written out as a fraction, and the numerator and denominator are multiplied by 1000 (not changing the value of the fraction), it becomes mg/l.

$$\mu g \times 1000 = mg$$

$$ml \times 1000 = l \qquad \text{Therefore: } \mu g/ml = mg/l$$

Milligrams per Kilogram

The unit label mg/kg is used widely within the environmental industry. It specifies the concentration of a solid within a solid, and is also a ppm designation, e.g., the concentration of copper in a chunk of gold would be mg/kg. The concentration of mercury in a fish would be mg/kg.

Molarity

This is the general chemist's preferred way to state concentration. Molarity is used as the concentration label in a wastewater laboratory when preparing acids and bases. It employs the gram molecular weight of the compound. A 1 molar (1 *M*) solution is the molecular weight of the chemical, in grams, dissolved in enough water to make 1 l.

Taking sodium hydroxide (NaOH) as an example, first calculate the molecular weight:

Na	23
O	16
H	1
	40

Therefore, 40 g NaOH dissolved in 1 l of water = 1 *M* NaOH.

Or, stated another way: A 1 *M* solution of NaOH = 40 g/l, or 40,000 mg/l.

Normality

Used widely in the water treatment industry, normality is similar to molarity. Instead of gram molecular weight, however, gram equivalent weight is used. The gram equivalent weight is the weight of a compound that contains 1 g atom available H or its equivalent. It is the number of grams of the compound that are required to react with 1 g H. To work this out simply, pick a compound:

NaOH or $Na^+ (OH)^-$. Count the charges on the positive ion. If it has one, then one molecule of it will react with 1 H ion (which also has one positive charge); it is 1 g equivalent. Therefore, for this compound, a 1 *M* solution = a 1 normal (*N*) solution.

40 g NaOH, dissolved in 1 l = 1 *M* = 1 *N*. However, if the compound has two or more H ion equivalents, then the *N* solution concentration is calculated based on each of those equivalents:

H_2SO_4 or $H_2^+ (SO_4)^-$. In sulfuric acid, there are two positive charges (two H ion equivalents). Calculate molecular weight:

2H	2
S	32
4O	+ 64
	98

98 g H_2SO_4 dissolved in 1 l = 1 *M* (1 g molecular weight)

98 g H_2SO_4 dissolved in 1 l = 2 *N* (2 g equivalent weights)

49 g H_2SO_4 dissolved in 1 l = 1 *N* (1 g equivalent weight)

1 *N* solution H_2SO_4 = 49 g/l or 49,000 mg/l

NOTE: Molarity is a matter of weights. Normality is a matter of weights and charges.

Liquids

In practice, however, these calculations are based on the assumption that the compound used is a dry chemical that can be weighed. HCl and H_2SO_4 are purchased in concentrated liquid form from the chemical distributor, and it is customary to measure a volume, not a weight. In addition, they are quite concentrated, and their liquid weight is greater per unit volume than water. Density is greater — specific gravity is greater.

For H_2SO_4, 1 ml is not the same weight as 1 g. One cannot measure out 49 ml and expect it to be 49 g. Concentrated sulfuric acid is 1.83 times as heavy as water (SG = 1.83).

To correct for this, divide the needed gram weight by the specific gravity.

$$49 \text{ g needed to prepare } 1\ N\ H_2SO_4.\ \frac{49}{1.83} = 27 \text{ ml}$$

Measure out 27 ml and fill to 1 l with water.

Add acid to water! Add acid to about half the volume of water in a beaker, mix, and fill to 1 l total. To obtain the correct specific gravity for a laboratory chemical, look on the bottle label, or check the *Handbook of Chemistry.* This type of calculation is needed in treatment plant use with liquid process chemicals, which are often shipped in concentrated form and are denser than water (liquid alum, liquid ferric chloride). The specific gravity should be calculated in when determining chemical dosage rates.

Dilution

There is one more factor to consider. In the laboratory, and for the treatment process, a purchased liquid chemical as shipped is already diluted in water. It is not 100% chemical. For instance, concentrated sulfuric acid is 96% acid in the chemical bottle. The other 4% is water. As calculated above, if this were 100% acid, 27 ml of it would be needed to prepare a 1 *N* solution. Since it is 96% pure, the required volume is $\frac{27}{.96} = 28$ ml. Concentrated HCl is only about 37% acid. In this case the difference is more significant.

Concentration of Common Reagents

	HCl	HNO_3	H_2SO_4	H_3PO_4	CH_3COOH	NH_4OH
Specific gravity	1.18	1.41	1.84	1.69	1.06	0.9
Strength of conc. reagent	37%	70%	96%	85%	100%	29%
Molecular weight	36.5	63	98.1	98	60	35.1
Molarity of conc. reagent	12	16	18	15	17	15
Number ml reagent to prepare 1 l of 1 *M* solution	83	64	56	70	42	66

NOTE: For preparation of common acids and bases, see the inside front cover of *Standard Methods*.

Percent Concentration

Percent by weight concentrations are used when dealing with concentrated chemical solutions and process sludges.

$$\frac{1 \text{ g chemical}}{100 \text{ g water}} = 1\% \quad \text{or} \quad \frac{1 \text{ g chemical}}{100 \text{ ml water}} = 1\%$$

$$\text{or} \quad \frac{10 \text{ g chemical}}{1000 \text{ ml water}} = 1\% \quad \text{or} \quad 10{,}000 \text{ mg/l} = 1\%$$

To convert a 1 *N* solution to percent concentration, first change it to mg/l, and then to percent.

$$1N\ H_2SO_4 = 49 \text{ g/l}$$

or 49,000 mg/l

or 4.9%

Changing Liquid Concentrations: (C × V = C × V)

The above formula is handy for dealing with laboratory chemicals or process chemicals. It can also be useful whenever liquid concentrations are changed. Any unit label for concentration or volume can be used. The formula originated from Boyle's and Charles' ideal gas law, PV = nRT. The pressure, temperature, and gas constant have been eliminated since they are the same for the solution before and after the concentration is changed. There are several different ways to use this formula:

The chemist has a 0.3 *N* solution of HCl. How many ml of it would be needed to prepare 100 ml of a 0.01 *N* solution?

Have this:		Want to get this:
C × V	=	C × V
0.3 *N* × V (ml)	=	0.01 *N* × 100 ml
V	=	3.3 ml

The chemist has 75 ml of a 1000-mg/l solution of NaCl. He wants to dilute this entire solution to a 50-mg/l solution. How much water will be needed for the dilution?

First calculate the amount of total solution that will result from making the dilution. This will also be needed to choose glassware size.

Have this:		Want to get this:
C × V	=	C × V
1000 mg/l × 75 ml	=	50 mg/l × V
1500 ml	=	V

Subtract original solution volume to determine water needed:

1500 ml total volume – 75 ml original NaCl volume = 1425 ml water needed.

For any chemical procedure, use common sense. If the volume indicated to make the dilution is too small to measure accurately with a pipet, make serial dilutions. Dilute it first to an intermediate strength, then down to the concentration needed.

Ion Concentration

At times it is desired to prepare a solution that has a specific concentration of a single element. The chemist must be able to calculate the amount of compound needed to obtain the correct concentration of the desired element.

It is desired to make up a solution that contains 100 mg/l of Na. The compound available is NaCl.

First, calculate the % weight of Na in this compound:

$$\begin{array}{ll} \text{Na} & 23 \\ \text{Cl} & \underline{+35} \\ & 58 \end{array} \qquad \frac{23}{58} = 0.4 = 40\% \quad \text{The compound is 40\% Na.}$$

Then divide the desired concentration by the % Na in this compound.

$$\frac{(100\text{ mg})}{0.4}\ (\text{sodium desired in the liter}) = 250\text{ mg}$$

Weigh out 250 mg NaCl. Dissolve in 1 l. It will be 100 mg/l Na.

This solution would be labeled 100 mg/l NaCl, as Na.

This same procedure can be used to calculate the concentration of any part of a molecule in a liquid solution. It is particularly useful with nitrogen ions, and it has become accepted to refer to these in terms of the weight of nitrogen (N) in the compound.

A solution that contains 100 mg/l $NaNO_3$, as N means that the compound in use is sodium nitrate, but 100 mg/l is the concentration of N in this solution.

Chapter 4

Laboratory Chemicals

Contents

The Chemical Label

Know your chemicals! When preparing reagents, always read the label on the chemical bottle. Anyone can make a mistake and pull the wrong chemical off the shelf. The chemical name is on the front of the bottle. Read it carefully. Some names that are very similar, e.g., sodium sulfate and sodium sulfite, are very different chemicals. The chemical state (powder, granular, liquid) is often listed under the name. If it is a liquid, the concentration is given. The exact chemical formula and the formula weight are listed on the left side. Be sure to compare them against the formula required for the test procedure. Under the formula is the assay. It gives a listing of impurities mixed with this chemical and their percentages. On the bottom of the label are special handling notes and warnings, such as Danger!

Poison! Some chemical distributors have adopted the National Fire Prevention Association safety hazard color codes for labeling and storing stock chemicals:

Gray: Moderate hazard

Blue: Health hazard; toxic if inhaled, ingested, or absorbed.

Yellow: Reactive, oxidizing agent

White: Corrosive

Red: Flammable

For most analyses, American Chemical Society (ACS) reagent grade chemicals are used. These are satisfactory except in certain instances where ultra-high purity chemicals are required. Any chemical that has been removed from the chemical bottle should be used or disposed of. Don't put it back into the bottle. To find out more about a chemical, consult the *Merck Index*, or the *Handbook of Chemistry. Standard Methods* also has useful information.

Chemicals and Water

Some dry chemicals are labeled "anhydrous." This means that the chemical formula is that of the compound only, e.g., NaCl may be labeled sodium chloride, or anhydrous sodium chloride, but either way is just NaCl.

Chemical Moisture

Some dry chemicals are hydroscopic and will readily absorb moisture from the air. This is just physical moisture and does not change the nature of the chemical or the formula. However, when preparing a solution using a specific amount of this chemical, some of what is being weighed out is the moisture. There is no way to know how much is moisture, and the prepared chemical solution will end up weaker than intended. This type of chemical must first be dried in an oven to drive off the water. Most laboratory procedures anticipate this problem, and include directions to dry certain chemicals for a stated amount of time before weighing. Do not exceed the required temperature, or the chemical may begin to decompose.

Bound Water

Some chemicals have chemically bound water attached to each molecule, e.g., manganous sulfate, $MnSO_4 \cdot 4H_2O$. This one may be labeled by the compound name only, or it may be labeled manganous sulfate-4 hydrate. As verified by the formula, this compound has four water molecules attached to every manganous sulfate molecule. This is not just a moist chemical. It is dry, but water is a permanent part of the molecule. Drying this chemical in an oven at normal chemical drying temperatures will not take this water off. Heating at much higher temperatures might, but then the compound would be decomposing. This is chemically bound water. It has weight, and that extra weight, if not accounted for, can throw the desired solution concentration off considerably. This is, however, a reliable, constant weight, and it can be calculated.

If the test procedure required preparing a solution with 10 g of $MnSO_4$, more than 10 g would be needed if the chemical chosen were $MnSO_4 \cdot 4H_2O$.

To calculate, first determine the percentage of $MnSO_4$ in this chemical. Add up the atomic weights of each element.

$MnSO_4 \cdot 4H_2O$		**$MnSO_4$**	
Mn	55	Mn	55
S	32	S	32
O	64	4O	64
8H	8		151
4O	64		
	223		

$\frac{151}{223} = 0.648 = 64.8\%$ of this chemical is $MnSO_4$. Now divide the weight needed by that percentage.

10 g needed: Therefore, 15.43 g of the chemical must be weighed out in order to get 10 g of $MnSO_4$.

Distilled Water

The quality of reagents is strongly dependent upon the quality of water used to prepare them. There must be no detectable concentration of the substance to be analyzed, or of anything which will react with that substance in the water. For use in a laboratory where multiple analyses are being performed, a source of the purest water available is needed: distilled water. Check its quality monthly by measuring the electrical conductivity (pure distilled water does not conduct electricity). EPA requires that distilled water in a certified water testing laboratory measure <1 μmho/cm at 25°C. However, contaminants do enter. The system is open to atmosphere; carbon dioxide and oxygen enter. If the collection bottle is not completely clean, nutrients will dissolve and algae will grow in the bottle, especially if it is near a window. As algae die, they provide food for bacterial growth. Dirty valves and delivery tubes are prime causes of sample and reagent contamination. Collection bottles should be acid washed periodically. They must be kept clean and closed. Distilled water does not maintain original purity over an extended period of time. It is best not to plan to store great quantities of it. It should be prepared as needed, and used.

> **NOTE: Check distilled water quality frequently. All prepared reagents depend on it.**

Chemical Indicators

Indicators are organic chemical dyes that react to produce a color under specific conditions. They are visual aids used to indicate the endpoint of titrations. Two types of indicators, absorption indicators and pH indicators, are most widely used in water laboratories.

Adsorption Indicators

When added to the sample, these chemicals loosely bind up a few molecules of the compound being measured, and a color change is noted. As the titration nears completion, the titrant demands these molecules back from the indicator and the color is discharged, marking the endpoint. For example, starch is an adsorption indicator used when titrating for dissolved oxygen. A few drops of starch produces a strong blue color in combination with the sample molecules. This color disappears immediately as those last molecules are taken by the titrant. This color change indicates that the titration is completed.

pH Indicators

A pH indicator changes color at a specific pH. These are used for titrations which are performed to neutralize an acid or alkaline component in the sample. Frequently used pH indicators are:

Phenolphthalein: Colorless (titrating with a base) → Pink: at pH 8.3

Methyl Orange: Orange (titrating with an acid) → Rose: at pH 4.5

How much to use? Directions for use of indicators usually specify "a few drops." This is a variable amount. Use enough indicator to obtain a decent color in which a change can be seen. For best visibility of end-point color change, titrate on a white surface. White paper under the flask works well.

> **NOTE: Chemical indicators are the only solutions in a water laboratory that are acceptable to measure with an eyedropper.**

Chemical Standards

A chemical standard is a solution that contains a known concentration of the chemical component of interest.

Use in Instrument Calibration

For example, a pH meter is calibrated to read correctly two solutions of known pH standards (for pH meters they are more commonly called "buffers"), before attempting to read the pH of the sample.

Use in Titration

A titrant must have an exact known concentration. It is the reagent that measures the component being tested for. A **primary standard** is a chemical solution that, as prepared, can be relied upon to be pure, and to be the exact concentration it is made up at. A **secondary standard** is a solution that is prepared from an impure chemical, and its exact concentration is not known. It is reacted with a primary standard (standardized) on a mole-to-mole basis, to determine its exact concentration. Most often it is the secondary standard that is the needed compound as a titrant for the test procedure. The purpose of the primary standard is to standardize the secondary standard.

Use in Colorimetric Analysis

These standards are prepared as a series of solutions with increasing known concentrations of the constituent to be analyzed. A reagent is added to each, which produces visible color, most often increasingly dense for the more concentrated standards. By adding the reagent to the unknown sample also, its color can be compared against those of the standards to determine concentration.

Stock, Standard, and Working Standards

In a colorimetric laboratory procedure that tests for very tiny amounts of a contaminant in a water sample, e.g., 2 mg/l expected, standards must be made up in very dilute concentrations. To minimize weighing problems and to increase shelf life of the standard, it is often practical to first prepare a much stronger stock solution, e.g., 1000 mg/l of the compound, and then dilute it down to a standard solution, e.g., 10 mg/l. From this standard solution the chemist chooses the working standards that will actually be used for the color comparison with the sample: 1 mg/l, 2 mg/l, 3 mg/l, and 4 mg/l. The working standards are prepared in a concentration range that will straddle the sample concentration and produce a linear slope. A prepared standard that has zero concentration of the substance to be analyzed is called a blank.

Chapter 5

Laboratory Instruments

Contents

Because a water laboratory is instrumental, it is important to understand some of the working mechanisms and principles upon which the instruments were developed and operate. No instrument functions perfectly forever. Proper maintenance, replacement of necessary components, and troubleshooting of some of the more obvious problems that occur are the chemist's responsibilities.

Measurement of pH

pH is defined as the negative log of the hydrogen ion concentration of the solution. This is a measure of the ionized hydrogen in solution. In short, it is the relative acidity or basicity of the solution. The chemical and physical properties, and the reactivity of almost every component in water are dependent upon pH. It relates to corrosivity, contaminant solubility, and the water's conductance, and has a secondary MCL range set at 6.5 to 8.5.

The pH Meter

pH measurement is probably the most widely used electrical method of analysis in a chemistry laboratory. This is a potentiometric method based on a voltage reading. The meter is a voltmeter and reads millivolts of electricity.

Working Mechanism

The sensing mechanism consists of two electrodes: a **reference electrode** and a **pH electrode**. Both must be immersed in the sample for a proper reading.

Reference Electrode

The reference electrode provides a constant voltage to the meter, which acts as a standard against which the sample reading is compared. The voltage is provided by a flow of electrolyte filling solution, a pH 7 saturated KCl solution (4 *M*), from the electrode into the sample through a small diffusion port near or at the bottom (notice a small, round, ceramic "clogged hole"). Fouling of this ceramic junction is the most frequent cause of electrode failure. Follow the manual cleaning procedure, or replace the electrode. The reference electrode is filled through a small opening near the top and must be kept full. The filling port has a sleeve or cap on it. It must be kept closed when not in use, to prevent evaporation of the electrolyte, and open when in use, allowing air flow.

Types of Reference Electrodes

A calomel reference electrode has Hg/HgCl internal wires. This type is used for most routine applications. It is easy to use and maintain, and uses saturated KCl electrolyte; shelf life is 2 to 3 years. A silver/silver chloride reference electrode has Ag/AgCl internal wires and is used for high temperature applications. It uses saturated KCl+AgCl electrolyte to prevent dissolving the internals. A double junction reference electrode is a special double-body type meant for use with very dirty samples, and with industrial samples that may cause interference due to a component in the sample. A different filling solution can be used. The gel-type reference electrode is made with a solid filling solution — a permanent gel not to be refilled. It does diffuse out into the sample slowly through the diffusion port just as the others do. The electrode is used until the gel is gone (approximately 1 year); the electrode is then replaced.

pH Electrode

This is a glass membrane electrode that actually senses the pH of the sample. Hydrogen ions diffuse through the membrane to the inside of the electrode where a wire picks up the reading and records the voltage carried in from the sample. This will be a different voltage reading than that coming in from the reference electrode. The meter records the difference between them in millivolts. It is then converted to pH.

Millivolts Compared to pH:			+		−	
	mv	7000	———	0	———	7000
	pH	0	———	7	———	14
			acid		base	

Combination Electrodes

A combination electrode is one that contains both electrodes in one probe. The reference electrode is the outer portion with filling port and bottom diffusion port. The pH electrode is the inner portion and extends all the way down to the very tip, which is a glass membrane. It is best to have an open plastic shield over the end of the probe to protect the fragile membrane from breakage. The advantage of the combination electrode is that only one probe is immersed in the sample. Polymer body combination electrodes (plastic), are made for rough handling applications, and are available in liquid-filled or gel-filled versions.

Meter Calibration

pH meters are calibrated (set to a correct reading) with purchased solutions that have specific pH values. They are called buffers, and are composed of phosphates, acetates, etc., depending upon the pH of the buffer. These are the standards with which the meter is set to work correctly. The pH is listed on each buffer bottle, and it is dependable, but shelf life is limited. Buffers can also be purchased in dry capsule form (long shelf life) and prepared as solutions when needed. Follow the instructions carefully on the meter for calibration. Instructions are somewhat different for each meter. Use the buffers indicated for calibration with that meter.

Always calibrate a pH meter with two buffers. A pH meter, near the end of calibration, will usually give a reading of "slope." This has to do with the efficiency of the electrode. It is the change in millivolts per decade of concentration. In pH terms, a decade of concentration is the difference between pH units, since the concentration multiplies by 10 from 1 pH unit to another. The instrument divides this change in mv/decade concentration by the ideal change in mv/decade concentration and obtains a percentage (of probe efficiency). The reading should be over 90% slope; some meters will adjust for the slight amount off 100% accuracy. If the slope reading is under 90%, repeat the calibration. If it is still off, recondition or change the electrode.

If you are using an older meter and there are no directions for calibration, choose two buffers that straddle the expected pH of the sample. Immerse the electrode in one of them and set the calibration knob to read that pH. Do the same with the other buffer, then set the meter to "read pH" and read the sample.

The buffer solution is really a "buffer." It neutralizes any possible acid or alkaline influences present in the water in which it is dissolved, and the pH should reside very stably at its designated value. When the pH electrode is immersed in the buffer, the reading should not drift. If it does, recondition or change the probe. A pH meter operates in millivolts, and it will give millivolt readings all the way through calibration, making the conversion to pH units at the end.

NOTE: Always calibrate a pH meter with two buffers.

pH Troubleshooting Tips

1. Open air port when using. Close when done.
2. Rinse between samples with distilled water.
3. Do not be concerned with odd readings obtained while the electrode is rinsing in distilled water. Although distilled water should have a pH of 7 at 25°C, the electrode has difficulty reading anything at all in distilled water, which does not conduct electricity.
4. Handle the pH electrode carefully. Glass membranes break easily.
5. Wait for the sample reading to settle down before recording it. Wait 5 to 6 min before taking a reading with any electrode method (cleaner waters often have low ionic strength and a slow response time).
6. If an automatic temperature adjustment probe is included with the meter, this must be immersed in the solution with the pH electrode in order to sense the temperature and read the correct pH.
7. Needle drift is frequently caused by the sample changing. Check for drift when immersed in pH 7 buffer. If there is drift when in the buffer, change the filling solution. Recondition or replace the electrode.
8. Don't stir the solution with the pH electrode. It may cause the needle to swing. However, if the sample is not stirred at all, a polarity may build up around the electrode. It is best to stir the sample very slowly with a stir bar.

9. Don't immerse the electrode in very high or low pH solutions. This shocks the electrode and it may have difficulty recovering.
10. Don't immerse the electrode in sample deeper than level of the electrolyte. Sample will flow in rather than electrolyte flowing out.
11. Electrode contamination from electrolyte crystallization can be rinsed off with distilled water. See conditioning instructions in manufacturer's pamphlet.
12. If readings are poor, it is more likely the electrode than the meter that is at fault.
13. Troubleshoot the obvious first:

    Is it plugged in?
    Is there electrolyte in the electrode?
    Is the meter switched to "pH"?
    Is the glass membrane broken?
    Is the sleeve on the filling port open?

14. After use, store electrode immersed in pH 7 buffer. Never let a pH electrode dry out.
15. Keep electrode clean between uses. pH 7 buffer is phosphate and it stimulates algae and bacterial growth. Change it frequently.

Ion-Specific Electrodes

Measurement by ion-specific electrode (ISE) is a widely accepted technology and is applicable to almost every laboratory and most dissolved ions. Its chief advantage is quick results. This is a rapid method of measuring for dissolved ions in the sample with a minimum of preparation time, reagents, space, and equipment. In addition, for most measurements, other substances or ions likely to be in the sample do not interfere with the reading. This is a definite advantage over colorimetric methods.

The chief disadvantage is cost. Most ISE electrodes have a usable life of 1 year and should be replaced annually.

These are potentiometric measurements. Any pH meter that reads out in millivolts can be used with ion-specific electrodes. The measurement of pH is actually an ion-specific electrode method; it is measuring hydrogen ions. In municipal and industrial water testing laboratories, it is common to see the measurement of nitrate, fluoride, chloride, and chlorine residual performed by ion-specific electrode. Since there are some differences from one electrode to another, more specific details are included in the test procedure chapters.

NOTE: For best results, replace ISE electrodes annually.

Colorimetry

Colorimetry began many years ago when chemists first realized that certain dissolved constituents in water, when added to specific reagents, will create a reproducible visible color, and that the density of color will be directly proportional to the concentration of that constituent in the water.

Beer's Law

In 1850, August Beer proposed the theory that the amount of light absorbed or transmitted through a colored solution will be directly proportional to the density of color of the solution. This is known as Beer's law, and it has been the basis of colorimetric analyses ever since. Johan Lambert also noted that the absorbance or transmittance of light through a colored sample is directly proportional to the width of the sample. This is Lambert's law. In modern colorimetric procedures, this factor is held steady by using the same size sample containers for all samples.

With this knowledge, the chemist was able to produce color in a normally colorless sample, in order to identify the presence of a contaminant and to distinguish between more or less concentrated contaminants. In 1852, for ammonia determinations in water samples (ammonia was at that time the prime indicator of polluted water), the use of Nessler Tubes was established. A series of increasingly concentrated standards was made up in test tubes, each containing a known amount of ammonia. When the color reagent was added, and color developed, the colors ranged from lighter to darker along the line of test tubes.The unknown sample, plus color reagent, developed a density of color that fell somewhere in among the standard colors, and its ammonia concentration was determined visually. With this, colorimetric methods were developed and the use of colorimetric standards was established.

Color Comparison Tubes

These were the first sets of colorimetric standards in widespread use. They are still considered adequate for rough comparison and are used in the field where instruments are not available. Permanent sets are available in a range of colors. When necessary, some chemists will make them up specifically to suit a project. Using permanent colors eliminates the task of making up a whole set of standards each time a test is run. There is, however, a risk of losing accuracy if the color hue of the standards is not comparable to that of the developed color in the sample, or if the color-producing reagent is not prepared in exactly the same way each time.

The Colorimeter

Improvements were made. To eliminate the tubes, the colorimeter was developed. Small, manually operated models are still found in some water treatment plants. The sample, with color-producing reagent added, is placed into a glass vial (cuvette), inserted into a small box, and held up to the light for viewing. A series of translucent glass or plastic colored discs arranged on cards or a color wheel, and labeled with their respective equivalent concentrations, can be held up alongside the sample to match the colors. Sets of different colors, reds, yellows, and blues for identification of separate constituents, could be purchased. Colorimeters gained popularity as a quick test method for many water constituents. Advantages over color comparison tubes included eliminating interference from varying daylight intensity and reducing equipment handling.

The major disadvantages of these early colorimetric methods were that the permanent color solutions or the discs did not always match the sample colors well, and the use of daylight, which is of different intensity each day and adds other colors into the sample, resulted in reduced accuracy. In addition, as each new batch of color-producing reagent was prepared, a slight change in its capacity to develop color would decrease the accuracy of the sample color. Finally, the human eye, a highly variable measuring instrument, was still used to read the colors.

The Spectrophotometer

By 1940, the automatic colorimeter, or spectrophotometer, had been invented. The source is an incandescent light that shines through a colored filter and deletes all wavelengths of light (all colors) except those most strongly absorbed by the sample's color. That portion of light which passes through the filter then passes on through the sample. Some of it is absorbed, but the light that is transmitted through the sample is sensed by a photocell, which converts it to a voltage. This is measured by a voltmeter and recorded on an analog scale (needle reading) as % absorbance or % transmittance. More recent improvements include a prism inserted just beyond the light source to break up the light into a rainbow of colors, separating all the wavelengths so that each color can be picked up individually. The filter is replaced with a movable opaque plate with a vertical slit in it. The chemist moves this plate back and forth, choosing the wavelength, until it resides in front of the exact wavelength of light he or she wishes to allow through to the sample. This provides maximum choice among visible wavelengths of light.

There may be several wavelengths of light that a given colored sample will absorb. The one at which it is absorbed most strongly will give a greater reading or "peak." For a given color, there may be two or more good peaks from which to work. The range of visible light on the electromagnetic wave spectrum is a small band from 400 to 700 nanometers (nm — the unit label for wavelength). It is within this range that cells in our eyes can pick up electromagnetic waves and register it to the brain as color. It is now possible to enter the ultraviolet or infrared zones with the spectrophotometer by using a different light source, one that will emit that type of light. Range, however, is limited because the instrument was designed for the visible light range; instrument materials and the surrounding air may interfere.

According to Beer's law, absorbance or transmittance of light can be sensed and recorded by the instrument. The spectrophotometer is first zeroed to "zero absorbance," or to 100% transmittance. Although absorbance mode is most frequently used in water chemistry, transmittance is used when the visible color is so slight that it will not absorb much light, or when the graphing of absorbance will be inverted.

Operation

For maximum accuracy, colorimetry has come full circle, back to the use of a set of standards prepared from the chemical constituent to be measured. To these, along with the sample, a color-producing reagent is added. Color is developed in all of them, and they are all read on the spectrophotometer. In this way all the colors are of the same hue, no interfering colors are given off from the light source, and a very precise and sensitive instrument does the reading.

Reagent Blank

The first step in operating a spectrophotometer is to zero the instrument. It is desired that the color sensed by the instrument be caused only by the developed color produced by the chemical being measured. Any color present in the reagent itself should be subtracted. To do this, prepare a blank made of distilled water plus the reagent and zero the instrument to this. Do this every time a test is run. It is really a standard with zero concentration of the measured constituent.

Calibration Curve

The instrument will yield a numerical reading for each of the standards and the sample, but that is a reading of % absorbance or % transmittance, on a scale of 0 to 1 or 1 to 10. It is not a concentration value for the standards or the sample. The concentrations of the standards are known, but to calculate the concentration of the sample a calibration curve is created from the standard readings, a slope is calculated, and from that slope the sample concentration is obtained.

Water utility personnel set up the graph like the one shown in Figure 5.1, with standard concentrations in mg/l across the bottom, and absorbance up the side. A dot or × is entered at the absorbance value for each standard after it is read on the instrument. A standard slope is then created by throwing a straight line midway between the dots which designate the standard readings. Often the dots do not fall exactly into a straight line. Do not draw dot to dot. Throw a straight line through them. This curve must be straight; it is an average. If the sample absorbance reading falls outside the range of standards, the test is officially invalid.

Calculating Sample Concentration

There are two methods to obtain sample concentration:

1. Create the calibration curve from the standard readings. Enter the sample reading manually from the left at its correct absorbance. Slide over to the curve, then straight down to concentration. This is the concentration of the sample. This is exactly how

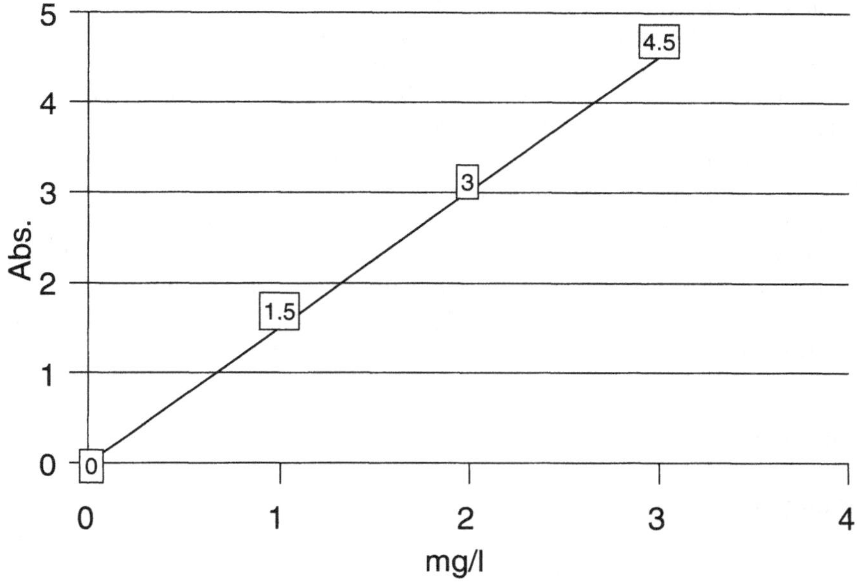

FIGURE 5.1
Calibration curve

sample and standard color were compared manually to Nessler tubes, that is, by holding the sample tube up against the standards to see which one matches. Using the instrument to make this comparison is much more accurate. Running the standards to prepare the calibration curve needs only to be done the first time the test is performed and each time a batch of fresh reagent is prepared, but the graph will have to be at hand to read sample concentration each time a sample is tested.

> **NOTE: EPA requires that at least three standards be run to prepare a calibration curve.**

2. After graphing the standards, calculate a numerical slope that will be representative of the slant of the standard curve: $\text{slope} = \dfrac{\text{x (mg/l)}}{\text{y (Abs.)}}$

Note that the formula is inverted from the familiar arithmetic slope (y/x). This doesn't matter. It still designates the slant of that line. This method came about because the water industry has chosen to place concentration on the bottom of the graph and absorbance up the side. Once the slope value has been calculated it will apply equally to the standards and the sample. Plug it back into the formula to calculate the concentration of the sample. For convenience, rearrange the formula to read:

$$\textbf{Sample Absorbance} \times \textbf{slope} = \textbf{Sample Concentration}$$

The first time the test procedure is run, prepare and run standards, sample, and reagent blank. Create calibration curve and calculate slope. After that, there is no need to refer to the graph each time the analysis is performed. Simply run sample and one standard to verify that the reading hasn't changed. Use the slope number to calculate concentration. The entire graphing procedure should also be done each time a new batch of reagent is prepared. A slight change in the nature of the reagent will change the graph and the slope number.

Many spectrophotometers sold today have much of this procedure built into the instrument. For a specific test, the standards have been previously entered into the unit and a standard slope created. To make use of this feature, turn the mode to "concentration" and just read in the sample. The instrument will do the standard comparison and provide results in concentration of sample.

Interferences

Turbidity in the sample interferes with colorimetric measurements. The spectrophotometer cannot distinguish turbidity from color, and a turbid sample will yield a false high reading. Filter the sample first with a 0.45-μm filter. In addition, a sample that is naturally **colored** will produce a false high reading. The color read by the instrument should originate only from the constituent/reagent complex, not from an extraneous natural color in the sample. Take a separate reading of the filtered sample with nothing added to it, and subtract this reading from the reading of the samples that have reagent added. Any color in the standards before the reagent is added also must be subtracted. To avoid this problem, choose a standard that has no color when dissolved.

Each colorimetric test procedure for chemical analysis has been developed for a specific range of concentrations. Accuracy is dependent upon obtaining a linear slope at that concentration range. Attempting to run a test using standard concentrations well out of the range for which the test was developed may produce a truly curved slope (which the graphing procedure attempts to force into a straight line). If the sample to be tested is out of range of the standard concentrations normally used for the test, it is better to dilute the sample than to raise the standard concentrations.

NOTE: Prepare standards in concentrations that are designed for the procedure.

Chapter 6

Quality Assurance/ Quality Control

Contents

A written and practiced Quality Assurance/Quality Control Program is now required at both water and wastewater treatment plants in the United States. Methods may vary somewhat from plant to plant, but the goal is to control and assure the quality of laboratory analyses.

When dealing with water samples, there is no written "right answer" that states how much turbidity, chlorine residual, or nitrate is in the sample. Chemical test results are used to control the treatment process, and these results are sent to the state for compliance reporting. The treatment plant laboratory cannot afford to be wrong. In a quality control program, definite statistical procedures and calculations are used to provide both precision and accuracy to laboratory work. This makes the data reliable.

Precision: Take a sample of water, split it into 50 equal aliquots, and test all 50 for nitrate. If exactly the same result is obtained on all of them, *that's precision.*

Accuracy: When doing this test, if the right result was also obtained with each of the 50 tests, *that's accuracy*. In testing water samples, strict adherence to standard methods, along with troubleshooting possible analysis and sampling problems, will provide reliability and the "right" result.

NOTE: The goal is to have the right result and to have it all the time.

Methods

Replicate Analyses: For **precision**

Run more than one sample through the entire test procedure. Differing results may show variation within a test method. This may apply to sampling or handling procedure, to changes within the sample, or to the instrument used. It will then take some troubleshooting to determine the cause.

Split Samples: For **accuracy**

Split a sample in two. Have one operator analyze one half and another operator analyze the other. Alternatively, share half with a neighboring plant. Either way, the results of both halves should be the same.

Reference Samples: For **accuracy**

Send to EPA or to the state laboratory for a sample that has been previously tested and has a known amount of the constituent in it. Test for that constituent. It should have the same result (sometimes these are referred to as reference standards).

Standard Samples: For **accuracy**

Make up a standard with a known amount of the constituent in it. Run this standard along with today's sample. If the test result from the standard is exactly the concentration that it was made up at, today's sample result is most likely correct also.

Spiked Samples: For **accuracy**

Add a known amount of the constituent to be measured for to the sample. Run one sample with the spike in it and one sample without the spike. For example, in testing the sample for chloride, add 1 mg/l chloride to one sample; the other sample is run with no addition. If the unspiked sample result is 4 mg/l chloride, then the spiked sample result has to be 5 mg/l chloride. If the spiked sample result is lower, then the test is only showing a percentage of the natural chloride concentration and spiked chloride concentration. The same applies if the result comes out higher. Differences may result from chemical interferences in the sample which would not show up by simply running a standard along with the sample, as in the method above.

Spiking a very low concentration sample is also useful to bring it up into the concentration range for which the test procedure was designed.

Armed with QA/QC data, one can then establish allowable deviations.

Percent Recovery

This statistical calculation is called **percent recovery of standard**. Using spiked samples, it can be determined what percent of the standard is recovered as a test result. 100% is complete recovery. **A range of 90 to 110% is considered acceptable.** Outside this range, errors should be corrected. For example, samples have been tested for fluoride on ten consecutive days. Each day an unspiked sample was run along with a sample spiked with 1 mg/l fluoride.

Plain Sample	Spiked Sample	Percent Recovery
1.02	1.93	91%
0.23	1.21	93%
0.03	0.96	93%
0.64	1.61	97%
2.01	3.01	100%
0.07	1.06	99%
0.03	1.01	98%
2.01	2.96	95%
1.01	1.93	92%
0.04	1.05	101%

$$\textbf{\% Recovery} = \frac{\textbf{Spiked Sample, mg/l} - \textbf{Plain Sample, mg/l}}{\textbf{Amount Spiked (mg/l)}}$$

Calculate the percent recoveries. How often do the results fall out of acceptable range? Average the percent recoveries for a month. Is the average acceptable? Determine where the results stand, statistically. This procedure checks accuracy of results. Done a number of times, it also checks precision. Percent recovery can be done on replicate samples, as well as on different samples. Recognition of an error is the first step. Continue to use the same statistical method to check accuracy as it improves until all results are in an acceptable range.

NOTE: These statistical methods only help determine whether and by how much something is wrong. They do not offer a solution to the problem.

Standard Deviation

Standard deviation is a statistical calculation that is done to check precision. It is defined as "variance from the mean." For example, take a group of 20 replicate samples tested for turbidity. All the results should have come out the same. If they don't, it first must be determined which is most likely the correct result. On a group of replicates like this, the best that can be done is to assume that the average of the results is most likely closest to the true answer. The larger the group of replicates done, the more assurance there is of this. This is assuming that accuracy of procedure is correct and only precision is off.

The **standard deviation** is the amount by which any one result varies from the average. The calculation assigns a number value to that amount of variance which is called one standard deviation (1S). If the sample being considered varies from the mean by no more than 1S (+ or –), then that sample result is acceptable. If the sample being considered varies from the mean by as much as 2S, steps should be taken to try to determine the cause of the sample variance. This is the "warning limit." If the sample being considered varies from the mean by as much as 3S, then this is bad data. 3S is the control limit.

Standard Deviation Formula

$$S = \sqrt{\frac{(\sum x^2) - \frac{(\sum x)^2}{n}}{n-1}}$$

Calculating Standard Deviation

Each Result: x

Number of Results:

Sum of Results: Σx^n

Sum of Results Squared: $(\Sigma x)^2$

Each Result Squared: x^2

Sum of Squared Results: Σx^2

Mean = Average of the Results = $\bar{x}$

A **control chart**, a graphical representation of the data and variances from the mean is constructed. Enter 1S, 2S, and 3S up and down the left side (+ and – from the mean. Enter data and construct the curve.

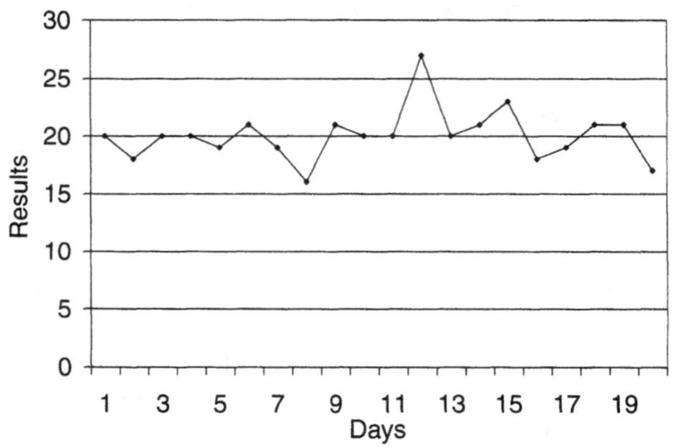

FIGURE 6.1
Control chart

Data				
	20 mg/l	21 mg/l	20 mg/l	18 mg/l
	18 mg/l	19 mg/l	27 mg/l	19 mg/l
	20 mg/l	16 mg/l	20 mg/l	21 mg/l
	20 mg/l	21 mg/l	21 mg/l	21 mg/l
	19 mg/l	20 mg/l	23 mg/l	17 mg/l

Reasons for Inaccuracies

There is inherent human error in everything we do. The challenge is to recognize errors and then to minimize them. The quality control procedures and calculations above help to recognize the errors. The following troubleshooting methods help to solve the problems once they are recognized. There are different kinds of errors:

Random Error

Random error is not consistent, and will vary from the true result by a random number. Examples of random error include:

- Muffle furnace whose temperature controller is not working well; temperature varies at one setting
- Volumes measured in beakers, using different ones each time
- Tests performed, samples taken, meters calibrated by different operators
- Regular instrument calibration neglected
- Titration endpoint called at different shades of pink by different operators
- Use of outdated reagents
- Use of dirty glassware
- Neglecting to shake the sample
- Neglecting to routinely check distilled water quality
- Neglecting to warm up the meter before use
- Neglecting to check temperature of incubators and ovens
- Neglecting to standardize reagents
- Neglecting to dry chemicals when called for
- Carelessness about keeping color development time the same for all colorimetric standards and samples before reading
- Rarely checking analytical balances with standard weights
- Neglecting to zero the analytical balance before each use
- Neglecting to mix reagents well enough in the volumetric flask before use
- Making volumetric measurements with hot liquids
- Performing a colorimetric procedure whose calibration curve is linear at 1 to 3 mg/l, with standards that range from 8 to 15 mg/l

- Keeping sulfites, nitrites, and arsenites for 2 months
- Using a light-sensitive reagent that has been stored on the windowsill
- Performing a chlorine residual test on a sample that has been sitting in the lab for 2 h

Systematic Error

Systematic error is consistent and varies from the true result by the same amount all the time. Examples of systematic error include:

- Using a distilled water blank instead of a reagent blank in colorimetric procedures
- Reading the meniscus on a pipet incorrectly
- Calibrating a meter incorrectly
- Using dirty optics on the spectrophotometer
- Measurement of viscous liquids and slurries in pipets and cylinders
- Using a pipet with a chipped tip that always delivers 0.4 ml less than its calibrated value
- Using a ruler with the end worn off
- Measuring 10 ml of liquid in a 100-ml graduated cylinder

> **NOTE: The waterworks industry measures in parts per million. Small inconsistencies can translate into large inaccuracies in analysis.**

Chapter 7

Total Alkalinity

Contents

Alkalinity is defined as the ability of water to resist a change in pH when acid is added; it relates to the pH buffering capacity of the water. Almost all natural waters have some alkalinity. This is nature's means of keeping the pH neutral, and stable, so that life can exist there.

Three major ions contribute to total alkalinity.

Hydroxide Ion: (OH^-)

Often referred to as "base," hydroxide ion will react with acid to yield water and salt:

$$(NaOH + HCl \rightarrow H_2O + NaCl)$$

Hydroxide ion has the greatest capacity of all the alkalinity ions to make the water basic. A pure water with only a few ions of hydroxide alkalinity will have a high pH. The hydroxide originates from the mineral bed in which the natural water body lies [$Ca(OH)_2$, $Mg(OH)_2$, $Fe(OH)_3$] and becomes soluble as acid components are added, making it available for reaction. However, because it does react rapidly with acidic components in the water, once they are neutralized, any hydroxide ion left in the water again forms insoluble precipitates. This is why hydroxide alkalinity is not often encountered in stable natural waters.

Carbonate Ion (CO_3^{2-})

Carbonates have a moderate capacity to raise the pH of water. Some carbonates are soluble in water. Some are insoluble, but they have great ability to neutralize acids. One carbonate ion can react with two hydrogen ions (from an acid):

Reacting with a strong acid: $Na_2CO_3 + 2HCl \rightarrow H_2CO_3 + 2NaCl$
The carbonate has changed two molecules of a strong acid to one molecule of a weak acid.

Reacting with a weak acid: $Na_2CO_3 + H_2CO_3 \rightarrow 2NaHCO_3$
The carbonate has changed the weak acid to an almost pH neutral compound, which itself will have the capacity to react with acid.

The amount of carbonate alkalinity in a water will depend upon the chemical components of the geologic strata upon which the water body rests, and their solubilities at various pHs.

Bicarbonate Ion (HCO_3^-)

In a pure water, bicarbonate has only slight capacity to raise the pH, even if there is an abundance of it. However, it has an excellent capacity to neutralize acids. For the most part, bicarbonate has originated from carbonates that have dissolved from the underlying rock, and have already reacted with acids, forming the bicarbonate.

Reacting with a strong acid: $NaHCO_3 + HCl \rightarrow H_2CO_3 + NaCl$
Bicarbonate changes one molecule of strong acid to a molecule of weak acid.

Reacting with a weak acid: $NaHCO_3 + H_2CO_3 \rightarrow H_2CO_3 + NaHCO_3$
No net change — both exist in equilibrium in the water.

In a natural water there is a balance; these ionic components are always in flux, always changing from one to another. Most natural waters have bicarbonate alkalinity only. The hydroxides and carbonates have already reacted with available acids and left the water with the stable bicarbonate.

Some natural waters have bicarbonate alkalinity and carbonate alkalinity. These two cannot react with each other to produce a different end product, and will exist in equilibrium.

> **NOTE: No natural waters can have all three: hydroxide, carbonate, and bicarbonate. The hydroxide reacts with the bicarbonate to form carbonate.**

The most common natural source of acid addition to a water is from CO_2 in the air. This gas is extremely soluble in water and ionizes to the weak carbonic acid (H_2CO_3). Notice its chemical components: actually, it is the acidic version of an alkalinity ion.

As CO_2 dissolves into the water, the water pH starts to drop; insoluble hydroxides and carbonates are made soluble. These alkalinity ions neutralize the acid and this brings more CO_2 into the water. In addition, algae in the water use up the CO_2, tending to raise the relative alkalinity (and pH) of the water. During algal blooms in a natural pond, as the pH rises the metal ions exceed their solubility limit, link with alkalinity ions, and start to drop out as precipitates, thus removing the alkalinity from the water and creating a drop in the pH. This is nature's balance. Temperature also has an effect. As the water warms, CO_2 becomes less soluble and starts to bubble out of the water. Remaining alkalinity ions will then have a greater effect on the pH.

Artificially added acids will complicate this balance, possibly upsetting it permanently. Acidic wastewater discharges and acid rain tend to drop the water's pH dramatically. In addition, if the rock strata that the natural water body lies upon is not laden with enough alkalinity compounds that can dissolve and neutralize the pH, it will remain low and severely limit the type of life that can exist in that water.

Alkalinity is a beneficial feature of a water, and does not necessarily indicate a high pH. The amount of total alkalinity is just as important as the type of alkalinity. A water with a few thousand molecules of bicarbonate alkalinity has a lot more buffering power than that same water with just a few molecules of hydroxide alkalinity, and yet the pH of the water remains close to neutral. Monitoring pH does not disclose much information about the amount of alkalinity in a water because basicity and alkalinity are not the same thing.

The three major alkalinity ions have been listed. Phosphates, borates, and others are also sources of alkalinity, but the concentrations of these in clean waters is so small that their effects are insignificant.

Treatment Plant Significance

Coagulant Addition

In potable water treatment, acidic metal compounds are used as coagulants. They have the capacity to create a hydroxide floc that will slowly settle and remove fine turbidity suspended in the water.

Metal coagulants are reactive enough to combine with the ions of the few water molecules that are disassociated. The floc settles and leaves the solution somewhat acid:

$$\underset{\text{alum}}{Al_2(SO_4)_3} + \underset{\text{water}}{6H^+ + 6OH^-} \rightarrow \underset{\text{floc}}{2Al(OH)_3} + \underset{\text{acid}}{3H_2SO_4}$$

However, if the pH of the water is allowed to drop significantly (below 5.5 in the case of alum), the floc will start to dissolve from the acid conditions and the chemical addition is ineffective. Therefore, it is important that there exists enough natural alkalinity in the water to buffer this chemical acid addition so that floc formation will be optimum, and the turbidity removal can proceed:

$$\underset{\text{alum}}{Al_2(SO_4)_3} + \underset{\text{alkalinity}}{3Ca(HCO_3)_2} \rightarrow \underset{\text{floc}}{2Al(OH)_3} + \underset{\text{soluble salt}}{3CaSO_4} + \underset{\text{carbon dioxide}}{6CO_2}$$

If there is not enough natural alkalinity in the water, some must be added. It is common to see that the utility adds some soda ash (Na_2CO_3) — carbonate alkalinity, along with the alum — to provide the alkalinity ions needed to react with the acid and keep the pH neutral:

$$\underset{\text{alum}}{Al_2(SO_4)_3} + \underset{\text{soda ash}}{3Na_2CO_3} + \underset{\text{water}}{3H_2O} \rightarrow \underset{\text{floc}}{2Al(OH)_3} + \underset{\text{soluble salt}}{3Na_2SO_4} + \underset{\text{carbon dioxide}}{3CO_2}$$

NOTE: Testing for total alkalinity in potable water treatment is most important for its relation to coagulant addition.

Water Softening

Chemical additives which are used to soften municipal water supplies combine with the natural alkalinity of the water to form precipitates. In this way, hardness-causing ions (calcium and magnesium) are made insoluble and removed from the water supply. Proper chemical dosage will depend on the amount and type of alkalinity in the water.

Corrosion Control

The presence of adequate alkalinity in a water supply neutralizes any acid tendencies, and prevents it from becoming corrosive. Monitoring tap water for alkalinity, calcium, orthophosphate, and silica may be required to initiate corrosion control if the water exceeds the lead/copper action levels.

Industrial

As industrial boiler water heats, carbon dioxide becomes insoluble and is removed with the steam. This leaves the water highly alkaline, often registering pHs greater than 10. If all of the calcium and other scale-forming ions were not previously removed from this water, they will combine with the alkalinity and precipitate onto the inner walls of the boiler, insulating it from the heat source.

Equivalent Weights and Ion Concentration in Potable Water Treatment

A special application of previously explained calculations is in use when dealing with hardness and alkalinity ions:

Alkalinity Ions:	OH^-	Hardness Ions:	Ca^{2+}
	CO_3^{2-}		Mg^{2+}
	HCO_3^-		

This concerns the ability of an alkalinity ion to react with acid, molecule for molecule. We often look at hardness ions in terms of their partner alkalinity ions. Therefore, for both of these, the reactive capacity is important, and for calculating concentrations of these ions, equivalent weights are used.

Ca^{2+}: molecular weight = 40; equivalent weight = 20

$CaCO_3$: molecular weight = 100; equivalent weight = 50

CO_2: molecular weight = 44; equivalent weight = 22

HCO_3^-: molecular weight = 61; equivalent weight = 61

Total alkalinity and total hardness may be each composed of several different ions:

Total alkalinity: OH^-, CO_3^{2-}, HCO_3^-

Total hardness: Ca^{2+}, Mg^{2+}

If measuring for each component separately, there is no question as to what unit label to attach. Measuring calcium: the result is 200 mg/l, as Ca^{2+}.

Yet when a measurement of the total amount is desired, a unit label is also needed, and the question arises: when measuring total hardness, the result is 250 mg/l, but is it calcium or magnesium, and how much of each?

Therefore, the most characteristic hardness/alkalinity compound has been chosen as the standard, $CaCO_3$, and unit labels for these constituents are most often registered "as if they were $CaCO_3$."

Another way to say this is, "If this were $CaCO_3$, how much would there be?"

A measurement of total alkalinity, total hardness, or any of the separate constituents of each of these can be mathematically converted to units of $CaCO_3$. This enables addition of them for a total, and it is very useful when calculating lime dosage for water softening. To do this, it is necessary to add several different kinds of units of alkalinity and hardness together.

To label in terms of $CaCO_3$: 200 mg/l Ca, as Ca^{2+} = 500 mg/l Ca, as $CaCO_3$.

To calculate:

Divide the concentration of the ion obtained by the percentage of the weight of $CaCO_3$, that the ion is.

Have 200 mg/l Ca, as Ca^{2+}.

Get percentage: MW Ca^{2+} = 40; Eq. wt = 20

MW $CaCO_3$ = 100; Eq. wt = 50

$$\frac{20}{50} = 0.4\ (40\%)$$

Then: $\frac{200}{0.4}$ mg/l Ca, as Ca^{2+} = 500 mg/l Ca, as $CaCO_3$

NOTE: Use equivalent weights in alkalinity calculations

Analysis: Titration Method for Total Alkalinity

A titration method is most commonly chosen to measure total alkalinity at potable water treatment plants. The test is performed on the raw water as well as on tap water.

A pH indicator that changes color at about pH 4.5 is added to the sample. An acid is titrated into the sample (usually 0.02 *N* H_2SO_4) until the pH drops to the indicator endpoint. At this point, all the alkalinity has been neutralized by the acid. The color change is noted and the titration is ended. The ml acid are recorded and a calculation of total alkalinity is made.

A pH meter may alternately be used in this method, titrating till the inflection point (sharp drop in reading) at about pH 4.5.

Quality Control

- Standardize acid titrant carefully.
- Titrate carefully. Don't overrun the endpoint.
- If using pH meter, warm it up for about 30 min before using.
- Calibrate meter with two pH buffers.

Apparatus

- Buret
- pH meter

Reagents

Sulfuric Acid

H_2SO_4, 0.02 *N*
The titrant
Dissolve 1.4 ml concentrated sulfuric acid in 1 l distilled water. This is approximately 0.05 *N* acid. *Must be standardized.*

Sodium Carbonate

Na_2SO_4, 0.05 *N*
Used to standardize the acid
Dry 3 to 5 g Na_2CO_3 at 250°C for 4 h and cool in a desiccator. Weigh 2.5 g, dissolve in distilled water, and fill to 1 l.

Methyl Orange

The 4.5 pH indicator
Dissolve 500 mg methyl orange powder in distilled water and dilute to 1 l. Color change is: Orange $\rightarrow$ Rose at pH 4.5.

Bromocresol Green

Alternate 4.5 pH indicator

Dissolve 100 mg bromocresol green, sodium salt, in 100 ml distilled water. Color change is: Blue → Yellow Green at pH 4.5.

Procedure

1. Standardize sulfuric acid:

   This procedure allows us to determine the exact normality of the prepared sulfuric acid and to adjust it to exactly 0.02 *N*.

   Using the pH meter, titrate the sulfuric acid into a small Ehrlenmeyer flask containing 50 ml Na_2CO_3. If the acid is exactly 0.05 *N*, then 50 ml of the titrant should bring the Na_2CO_3 solution to the pH 4.5 inflection point (neutralized).

   However, both solutions were dissolved in distilled water, which has CO_2 (an acid) in it. To eliminate the CO_2, proceed with the titration down to pH 5. Then boil the solution for 5 min under a watch glass. Cool the solution (pH should go up a little).

   Continue the titration down to the inflection point; record total ml acid used.

   Calculate: $C \times V = C \times V$ for the true normality of the acid.

   acid carbonate

   Dilute the acid down to 0.02 *N* for use in the total alkalinity test.

2. Measure 100-ml sample in a 250-ml Erlenmeyer flask. Add a few drops of methyl orange or bromocresol green indicator and mix.
3. Titrate 0.02 *N* sulfuric acid into sample, swirling flask, until endpoint color change is noted. Record ml titrated.
4. Calculate: $$\textbf{mg/l Total Alkalinity, as CaCO}_3 = \frac{\textbf{ml titrated} \times \textbf{1000 ml/l}}{\textbf{ml sample}}$$

   If a different normality acid is used for this titration, a longer version of the formula will adapt to this:

$$\text{mg/l Tot. Alk., as CaCO}_3 = \frac{\text{ml titrant} \times \text{Norm. titrant} \times \text{equiv. wt. CaCO}_3 \times 1000 \text{ ml/l}}{\text{ml sample}}$$

Federal Limits

None.

References

American Public Health Association, American Water Works Association, and Water Environment Federation, *Standard Methods for the Examination of Water and Wastewater*, 20th ed., American Public Health Association, Washington, D.C., 1998, p. 2-26.

American Water Works Association, *Basic Science Concepts and Applications*, 2nd ed., American Water Works Association, Denver, CO, 1995, p. 427.

Analysis: Titration Method for Distinguishing Various Types of Alkalinity Ions

If desired, a separation of the three different types of alkalinity ions (OH^-, CO_3^{2+}, HCO_3^-) can be chemically determined. This at times is done:

a. If there has been a noticeable and unusual change in the total alkalinity in the source water
b. If contamination of the source is suspected
c. For operation of a water softening plant; the type of alkalinity is important in calculation of chemical dosage

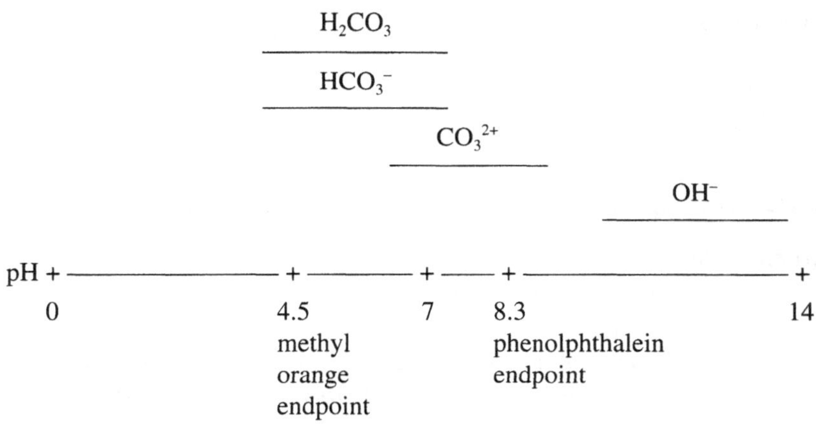

Relative to a pH scale, the various types of alkalinities reside as in the diagram above. In other words, these ions have the capacity to hold the pH of the water at these ranges, depending on how much of each is present. Notice the overlap between carbonate alkalinity and bicarbonate alkalinity. Now picture titrating with an acid into a water with carbonate alkalinity. By the time pH 8.3 is reached, exactly half the carbonate alkalinity has been neutralized. This is always the case. One of the reasons that phenolphthalein was chosen as a useful indicator for this method was that it does visually show this point.

To explain more accurately, in titrating down to pH 8.3, all the carbonate alkalinity has been changed to bicarbonate alkalinity. Half of its buffering power of this ion has been neutralized; the other half is still there, as bicarbonate.

If using a pH meter, an inflection point (sudden drop) is noted at this point also, as all the carbonate alkalinity is half neutralized.

> **Note:** **As each form of alkalinity is neutralized by the acid it doesn't disappear, but changes to the next "lower on the pH scale" form of alkalinity.**

Carbonic acid is included in the above line. It is not considered true alkalinity, but is the weak acidic version of it, to which the alkalinity ions are converted when reacting with a stronger acid.

When titrating to separate the types of alkalinity in the sample, acid is titrated to pH 8.3 color change using phenolphthalein. Total alkalinity indicator is added, and titration continues to color change at pH 4.5.

Quality Control

- Standardize acid titrant carefully.
- Titrate carefully. Don't overrun the endpoint.
- If using pH meter, warm it up for about 30 min before using.
- Calibrate meter with two pH buffers.

Apparatus

- Buret
- pH meter

Reagents

Sulfuric Acid

H_2SO_4, 0.02 *N*
The titrant
Dissolve 1.4 ml concentrated sulfuric acid in 1 l distilled water. This is approximately 0.05 *N* acid. *Must be standardized.*

Sodium Carbonate

Na_2SO_4, 0.05 *N*
Used to standardize the acid
Dry 3 to 5 g Na_2CO_3 at 250°C for 4 h and cool in a desiccator. Weigh 2.5 g, dissolve in distilled water, and fill to 1 l.

Methyl Orange

The 4.5 pH indicator
Dissolve 500 mg methyl orange powder in distilled water and dilute to 1 l. Color change is: Orange → Rose at pH 4.5.

Bromocresol Green

Alternate 4.5 pH indicator
Dissolve 100 mg bromocresol green, sodium salt, in 100 ml distilled water. Color change is: Blue → Yellow Green at pH 4.5.

Procedure

1. Standardize sulfuric acid:

   This procedure allows determination of the exact normality of the prepared sulfuric acid and adjustment of it to exactly 0.02 *N*.

   With 50 ml Na_2CO_3 in a small Erlenmeyer flask, titrate the sulfuric acid into it, using the pH meter. If the acid is exactly 0.05 *N*, then 50 ml of the titrant should bring the Na_2CO_3 solution to the pH 4.5 inflection point (neutralized).

   However, both solutions were dissolved in distilled water, which has CO_2 (an acid) in it. To eliminate the CO_2, proceed with the titration down to pH 5. Then boil the solution for 5 min under a watch glass. Cool the solution (pH should go up a little).

   Continue the titration down to the inflection point; record total ml acid used.

   Calculate: $C \times V = C \times V$ for the true normality of the acid.

   acid carbonate

   Dilute the acid down to 0.02 *N* for use in the total alkalinity test.

2. Measure 100-ml sample in a 250-ml Erlenmeyer flask. Add a few drops of phenolphthalein indicator and mix.
3. If sample shows pink color: Titrate 0.02 *N* sulfuric acid into sample, swirling flask until endpoint color change is noted (pink → colorless) Stop titration. Record ml titrated.
4. Add a few drops of methyl orange or bromocresol green indicator and mix.
5. Titrate 0.02 *N* sulfuric acid into sample, swirling flask until endpoint color change is noted. Record ml titrated.
6. Calculate:

$$\textbf{mg/l Carbonate Alkalinity, as CaCO}_3 = \frac{\textbf{ml titrated for the carbonate} \times \textbf{1000 m/l}}{\textbf{ml sample}}$$

Explanation of Alkalinity Ions Separation

If the water was:

Sample 1:
This sample has hydroxide alkalinity only.

- Initial pH would be over 10.
- Pink color visible when phenolphthalein is added.
- Large inflection point at 8.3; sudden drop to 4.5 and below.

Sample 2:
This sample has carbonate alkalinity only.

- Initial pH is over 8.3.
- Pink color visible when phenolphthalein is added.
- Half the acid ml is titrated to phenolphthalein endpoint.
- Half the acid ml is titrated below phenolphthalein endpoint.

Sample 3:
This sample has hydroxide plus carbonate alkalinity.

- Initial pH is over 8.3.
- Pink color when phenolphthalein is added.
- More than half the acid ml is titrated in getting *to* the phenolphthalein endpoint. This will include all the OH^- alkalinity and half the carbonate alkalinity.
- The other half of the carbonate alkalinity is titrated below the phenolphthalein endpoint.

Sample 4:
This sample has carbonate plus bicarbonate alkalinity.

- Initial pH is over 8.3.
- Pink color when phenolphthalein is added.
- Less than half the acid ml is titrated to the phenolphthalein endpoint. This will be half the carbonate alkalinity.
- More than half the acid ml is titrated below the phenolphthalein endpoint. This includes the other half of the carbonate, plus the bicarbonate alkalinity.

Sample 5:
This sample has bicarbonate alkalinity only.

- Initial pH is under 8.3.
- No color when phenolphthalein is added.
- Titrate down to methyl orange endpoint; it's all bicarbonate.

Federal Limits

None.

References

American Public Health Association, American Water Works Association, and Water Environment Federation, *Standard Methods for the Examination of Water and Wastewater*, 20th ed., American Public Health Association, Washington, D.C., 1998, p. 2-26.

American Water Works Association, *Basic Science Concepts and Applications*, 2nd ed., American Water Works Association, Denver, CO, 1995, p. 427.

Chapter 8

Total Hardness

Contents

Hardness is defined as the sum of the concentrations of calcium and magnesium ions dissolved in water. These two ions are the major hardness constituents, and though some other metals contribute to hardness, their concentrations in natural waters are so much smaller that their significance as hardness is negligible.

Calcium is the most abundant dissolved cationic constituent of natural fresh waters, and is widely distributed in the minerals of rocks and soils. It is the fifth most abundant element on earth and is found in every major land area of the world.

Magnesium is also a major constituent of rocks, in abundance second to calcium (about one fifth as much), and is usually found in occurrence with calcium.

The carbonate salts of calcium are the major source of dissolved calcium and are generally referred to as limestone or calcite. They include Iceland spar (pure), marble and alabaster (less pure and more compressed), and chalk. They can be white or colorless. Calcium carbonate is quite insoluble in pure water and dissolves only up to 15 mg/l, but if CO_2 is present, this natural acidity makes the limestone much more soluble. This often occurs in groundwaters; bacterial action in the soil releases CO_2, changing the carbonate to bicarbonate, and dissolving large amounts of calcium into the water. Because of this, groundwaters are generally harder waters than surface waters.

To the consumer, a more appropriate definition of hardness is the water's effect on scaling, corrosion, and soap. With hard water, it is difficult to produce a soap lather. Hard waters leave spots on glasses, dingy film on laundry and hair, and crusty deposits on bathroom fixtures.The presence of hardness in water supplies contributes no taste, odor, color, or turbidity to the water. Water hardness is purely an aesthetic property, and has no health significance. It varies considerably from place to place in this country, and a public water utility may or may not decide to remove it from the supply, depending upon the concentration and stability of the hardness, and the economics of the community. Water softening is a costly operation and one that is not easy to optimize. Most communities leave the water softening up to the individual consumer.

Generally, a water with a total hardness over 200 mg/l, as $CaCO_3$, would be considered hard water.

NOTE: Consumers who have ion exchange water softeners installed in their homes may identify hardness in terms of "grains per gallon." There is a direct conversion: 17.1 mg/l = 1 grain per gal.

Treatment Plant Significance

The test for total hardness is routinely performed at most public water utilities, whether or not they soften the water.

Scaling

Monitoring hardness aids in predicting whether the water will form scale in pipes or boilers and helps the utility to determine the need and amount of sequestering chemicals to alleviate scale-forming tendency.

Water Softening

Measurement of calcium and magnesium are needed to determine chemical dosage for lime/soda ash softening plants and the size of ion exchange softening units.

Types of Hardness

In Terms of the Metal Ion

Calcium Hardness: Concentration of calcium is routinely measured separately from total hardness. Its concentration in waters can range from 0 to several thousand mg/l, as $CaCO_3$.

Magnesium Hardness: Magnesium is routinely determined by subtracting calcium hardness from total hardness. There is usually less magnesium than calcium in natural water. Lime dosage for water softening operation is partly based on the concentration of magnesium hardness in the water.

In Terms of the Anions Associated with the Metal Ions

Carbonate Hardness: The hardness ions associated with carbonates and bicarbonates ($CaCO_3$, $Ca(HCO_3)_2$, $MgCO_3$, $Mg(HCO_3)_2$). A consumer term, **temporary hardness**, was formerly used for this type of hardness (it precipitates when the water was boiled, and therefore, is temporary). It is of interest to all utilities to keep track of how much of the hardness is carbonate hardness, for this is the component that is associated with scaling tendency in distribution piping. It relates to the amount of alkalinity in the water.

NOTE: Carbonate hardness is hardness combined with the water's alkalinity.

Noncarbonate Hardness: The hardness ions associated with other anions ($CaSO_4$, $CaCl_2$, $MgSO_4$, $MgCl_2$). Previously termed **permanent hardness** because it could not be made to precipitate by boiling the water.

The major significance in separating carbonate from noncarbonate hardness is for water-softening calculations. Lime and soda ash dosage will depend upon the concentrations of each of the two types of hardness.

NOTE: If the total hardness is greater than the total alkalinity, then the difference between them is the noncarbonate hardness. (TH – TA = NCH)

Analysis: Titration Methods for Total and Calcium Hardness

Total Hardness: This is a simple titration which makes use of an organic adsorption indicator to define the endpoint. The indicator, Eriochrome Black T (viscous, dark blue color) is added to the sample. It loosely binds up a few calcium and magnesium ions, creating a calcium or magnesium complex compound that exhibits a wine red color. The reagent is then titrated into the sample. The sodium salt of ethylenediamine tetraacetate (EDTA) is the titrant, a widely used compound in laboratory and industry which has chelating properties. It immediately ties up the free calcium and magnesium ions. When there are no more free ions left, the titrant combines with the last few that are loosely bound to the indicator. It then turns back to the blue color, and the endpoint is reached. An ammonium buffer is added to raise the pH, speed up the reaction, and sharpen the endpoint.

Calcium Hardness: This principle of this test is the same as that for the total hardness test. A different indicator is used, and sodium hydroxide is used as the buffer.

Quality Control

- Standardize EDTA titrant.
- Titrate carefully and quickly. The buffer raises the solution to pH 10, and hardness ions may start to precipitate out. Titration must be completed before this happens. Insoluble calcium will not be included in the results of this test.

Apparatus

- Buret

Reagents

Test for Total Hardness:

Buffer Solution

Sharpens the endpoint

Dissolve 16.9 g ammonium chloride in 143 ml concentrated ammonium hydroxide. Add 1.25 g magnesium salt of EDTA and dilute to 250 ml with distilled water.

Eriochrome Black T

Titration indicator

Dissolve 0.5 g Eriochrome Black T in 100 g triethanolamine. Solution will be dark blue and thick. Store in small eyedropper bottle.

EDTA

0.01 *M*

The titrant

Weigh 3.723 g disodium salt of EDTA, dissolve in distilled water, and dilute to 1 l. *Must be standardized.*

Standard Calcium Solution

1000 mg/l

Primary standard to standardize the titrant: Weigh 1 g anhydrous $CaCO_3$ powder; put in a 500-ml Erlenmeyer flask. Place funnel in the flask neck and add, a little at a time, a solution of concentrated HCl diluted in half with distilled water until all $CaCO_3$ is dissolved. Add 200 ml distilled water and boil for a few minutes to expel CO_2. Cool, add a few drops of methyl red indicator, and adjust to orange color by adding 3 *N* NH_4OH or HCl solution diluted in half with distilled water. Dilute to 1 l.

Ammonium Hydroxide

NH_4OH, 3 *N*
To prepare standard calcium solution
Dilute 200 ml concentrated NH_4OH to 1 l.

Methyl Red

Indicator used to prepare standard calcium solution
Dissolve 100 mg methyl red sodium salt in distilled water and dilute to 100 ml.

Procedure

1. Standardize EDTA:

   Dilute 25 ml standard calcium solution to 50 ml with distilled water. Add 1 ml buffer, then 1 to 2 drops indicator; mix. Solution will turn red. Titrate EDTA into solution until red color is gone (turns blue). Record ml used.

   Calculate: $\text{Hardness (EDTA), mg/l, as CaCO}_3 = \dfrac{\text{ml titrated} \times 1000\ \text{ml/l}}{\text{ml sample}}$

   Where: A = mg $CaCO_3$ equiv. to 1 ml EDTA
2. Measure 50 or 100 ml sample, depending on expected hardness. Add 1 to 2 ml buffer (to reach pH 10). Add 1 to 2 drops indicator; mix. Solution will turn red.
3. Titrate EDTA titrant into solution until red color is gone (turns blue). Record ml used.
4. Calculate:

$$\textbf{Total Hardness, mg/l, as CaCO}_3 = \frac{\textbf{ml titrated} \times \textbf{1000 ml/l}}{\textbf{ml sample}}$$

Test for Calcium Hardness:

Buffer Solution

NaOH, 1 *N*
Sharpens the endpoint
Measure 40 g NaOH, dissolve in distilled water, and dilute to 1 l.

Murexide

Titration indicator

Dissolve 150 mg murexide (ammonium purpurate) in 100 g absolute ethylene glycol. Solution is stable for 1 day.

Alternately, prepare the stable dry version. Mix 200 mg murexide with 100 mg NaCl. Grind to 50 mesh.

Absolute Ethylene Glycol

For indicator preparation

Sodium Chloride

NaCl

For indicator preparation

EDTA

0.01 *M*

The titrant

Weigh 3.723 g disodium salt of EDTA, dissolve in distilled water, and dilute to 1 l. Must be standardized.

Standard Calcium Solution

1000 mg/l

Weigh 1 g anhydrous $CaCO_3$ powder and put in a 500-ml Erlenmeyer flask. Place funnel in the flask neck and add, a little at a time, a solution of concentrated HCl diluted in half with distilled water until all $CaCO_3$ is dissolved. Add 200 ml distilled water and boil for a few minutes to expel CO_2. Cool, add a few drops of methyl red indicator, and adjust to orange color by adding 3 *N* NH_4OH or HCl solution diluted in half with distilled water. Dilute to 1 l.

Ammonium Hydroxide

3 *N*

To prepare standard calcium solution

Dilute 200 ml concentrated NH_4OH to 1 l.

Methyl Red

Indicator used to prepare standard calcium solution

Dissolve 100 mg methyl red sodium salt in distilled water and dilute to 100 ml.

Procedure

1. Standardize EDTA. See procedure for EDTA standardization in Total Hardness procedure.
2. Measure 50- or 100-ml sample depending on expected hardness. Add 1 to 2 ml buffer to reach pH 10. Add 1 to 2 drops indicator and mix. Solution will turn pink.
3. Titrate EDTA titrant into solution until pink color is gone (turns pale purple). Record ml used.
4. Calculate:

$$\textbf{Calcium Hardness, mg/l, as } \mathbf{CaCO_3} = \frac{\textbf{ml titrated} \times \textbf{1000 ml/l}}{\textbf{ml sample}}$$

Federal Limits

None.

References

American Public Health Association, American Water Works Association, and Water Environment Federation, *Standard Methods for the Examination of Water and Wastewater*, 20th ed., American Public Health Association, Washington, D.C., 1998, pp. 2-37, 3-64.

American Water Works Association, *Basic Science Concepts and Applications*, 2nd ed., American Water Works Association, Denver, CO, 1995, p. 413.

Chapter 9

Acidity

Contents

Carbon dioxide, the major acid component of all natural waters, enters surface waters by absorption from the atmosphere, and from aerobic and anaerobic bacterial action in the water. It is very soluble (up to 2000 ppm) and is always entering and leaving solution, depending on its partial pressure with atmospheric CO_2. Once in the water, CO_2 is in equilibrium with its ionized form, H_2CO_3 (carbonic acid), is being utilized by algae and other plant life, and reacting with the natural alkalinity of the water. Carbon dioxide is always in flux, and in clean surface waters 10 to 25 mg/l, as $CaCO_3$, is a normal concentration.

In groundwater, however, the concentration may be higher. Water percolating through the ground is acted upon by bacteria, aerobic near the surface and anaerobic below; the CO_2 released by their action is locked in. If the soils involved do not contain much calcium or magnesium carbonate to neutralize this acid, the water may permanently be left with a greater concentration of CO_2. The water may be corrosive; pH may be depressed and theoretically down to pH 4.5.

It is this CO_2 acidity that is monitored in the acidity test, and it can exist in natural waters which are under a pH of 8.3 (phenolphthalein endpoint). It is of no direct health significance in drinking waters. Our natural stomach acid is concentrated

HCl, with a pH between 0 and 1. The value of monitoring CO_2 content is in recognizing its capacity to react with other chemicals in the water.

NOTE: The acidity test measures CO_2 concentration.

Treatment Plant Significance

Effect on pH

Carbon dioxide acidity, if high, will depress the pH of the water and may influence the effectiveness of a metal coagulant added to remove turbidity.

Water Softening

Carbon dioxide reacts with lime used to soften water. Lime dosage must take into account the amount of CO_2 in the water.

Analysis: Titration Method for Water Acidity

Carbon dioxide acidity exists between pH 4.5 and 8.3 in equilibrium with bicarbonate alkalinity. Where the pH is below 4.5 (methyl orange endpoint), mineral acidity exists. It would be unusual to have mineral acidity in a clean drinking water source.

Quality Control

- Standardize NaOH titrant.
- Pour aliquot into flask carefully. Avoid shaking.
- Calibrate pH meter with two buffers.

Apparatus

- pH meter
- Buret

Reagents

Sodium Hydroxide

NaOH, 0.1 *N*
The titrant
Dissolve 4 g NaOH in 1000 ml distilled water. *Must be standardized.*

Potassium Hydrogen Phthalate

$KHC_8H_4O_4$, 0.05 *N*

To standardize the titrant

Crush 20 g $KHC_8H_4O_4$ and dry at 120°C for 2 h. Cool, weigh 10 g, and dilute to 1000 ml.

Bromocresol Green

4.5 pH indicator

Dissolve 100 mg bromocresol green, sodium salt, in 100 ml distilled water.

Color change is: Yellow Green → Blue at pH 4.5.

Methyl Orange

Alternate 4.5 pH indicator

Dissolve 500 mg methyl orange powder in distilled water and dilute to 1 l. Color change is: Rose → Orange at pH 4.5.

Phenolphthalein

8.3 pH indicator

Dissolve 5 g phenolphthalein in 500 ml 95% ethyl alcohol and add 500 ml distilled water.

Carbon Dioxide Free Water

Prepare all solutions with this. Boil distilled water for 15 min and cool to room temperature. The pH of this water should be 6, and conductivity should be <2 µmhos/cm.

Procedure

1. Standardize NaOH titrant:

   Measure 40 ml $KHC_8H_4O_4$ into an Erlenmeyer flask. Titrate with 0.1 *N* NaOH using a pH probe, to the inflection point at about pH 8.7. Calculate NaOH normality:

$$\text{Normality} = \frac{A \times B}{204.2 \times C}$$

   A = g $KHC_8H_4O_4$ weighed into a 1-l flask

   B = ml $KHC_8H_4O_4$ taken for titration

   C = ml NaOH titrated

   Dilute the NaOH to 0.02 *N*.

2. Measure pH. If below 4.5, indicating mineral acid present, add methyl orange or bromocresol green, and titrate with 0.02 *N* NaOH until the color change occurs.
3. Add phenolphthalein and continue titration until the pH 8.3 endpoint is reached. Record ml used in titration from pH 4.5 to 8.3.
4. Calculate: $$\textbf{Acidity, as } \mathbf{CaCO_3} = \frac{\textbf{ml titrated} \times \mathbf{1000}}{\textbf{ml sample}}$$

Federal Limits

None.

Reference

American Public Health Association, American Water Works Association, and Water Environment Federation, *Standard Methods for the Examination of Water and Wastewater*, 20th ed., American Public Health Association, Washington, D.C., 1998, p. 2-24.

Chapter 10

Water Softening — Jar Test for Chemical Dosage

Contents

The lime/soda ash method is the preferred method used by municipal utilities for softening drinking water supplies.

Lime (CaO, quicklime) reacts with water (is slaked) and becomes $Ca(OH)_2$ (slaked lime, or hydrated lime). Still mostly insoluble in stock concentrations, it is then carried to the process as a slurry and dissolved into the flow.

Soda ash (Na_2CO_3) may also be added for removing the hardness that the lime cannot precipitate. It is soluble in water and relatively easy to feed into the flow.

Alternate Methods of Water Softening

For some industrial purposes, it is necessary to remove most or all hardness from the water supply. Frequently used are lime/soda ash softening, often under heat and pressure for greater efficiency and silica removal, followed by ion exchange softening.

Ion exchange softening removes all hardness, but leaves all the alkalinity in the water. This is a cation exchange process. If it is desired to remove the alkalinity as well, anion exchange can also be used. Membrane processes, reverse osmosis, and nanofiltration provide removal of all hardness ions and dissolved organics, and are now employed by industry and municipalities.

Treatment Plant Significance

The operation of a lime/soda ash softening plant raises the pH of the water enough to shift the bicarbonate alkalinity to the carbonate and hydroxide forms. Hence, these ions will form insoluble compounds with the hardness ions and precipitate out of the water. Lime adds the hydroxide alkalinity that achieves this. Soda ash is used to remove the noncarbonate hardness. When this is present there isn't enough alkalinity for all the hardness ions to combine with to precipitate an insoluble compound. The soda ash adds that alkalinity.

Calculation of chemical dosage is done first mathematically, then with jar testing, before being applied to full-scale treatment.

The Reactions

Lime is used to remove carbonate hardness. It reacts with:

(1) Carbon dioxide: $CO_2 + Ca(OH)_2 \rightarrow \underset{\text{insoluble}}{CaCO_3} + H_2O$

(2) Calcium bicarbonate: $Ca(HCO_3)_2 + Ca(OH)_2 \rightarrow \underset{\text{insoluble}}{2CaCO_3} + 2H_2O$

(3) Magnesium bicarbonate: $Mg(HCO_3)_2 + Ca(OH)_2 \rightarrow \underset{\text{soluble}}{MgCO_2} + 2H_2O$

(4) Magnesium carbonate: $MgCO_3 + Ca(OH)_2 \rightarrow \underset{\text{insoluble}}{Mg(OH)_2}$

Referring to Eq. 2, lime reacts with soluble calcium bicarbonate to form insoluble calcium carbonate. It is not necessary to add lime to remove carbonates. Calcium carbonate is insoluble, and won't exist dissolved in the water in concentrations of more than a few parts per million. There is no need to remove carbonates of other positively charged ions. These are not hardness ions, e.g., Na_2CO_3.

If the total alkalinity is greater than the total hardness, the concentration of the total hardness, instead of the bicarbonate alkalinity, is used to guide lime dosage. There is no need to add lime for a concentration of alkalinity that has no hardness to precipitate out.

Referring to Eq. 3, note that the magnesium carbonate product of the reaction is fairly soluble. Nothing has been removed here yet. Extra lime is needed to actually precipitate the magnesium. For this reason, the magnesium concentration is added to the dosage calculation separately.

Soda ash alone is used to remove calcium noncarbonate hardness.

$$CaSO_4 + Na_2CO_3 \rightarrow \underset{\text{insoluble}}{CaCO_3} + Na_2SO_4$$

Lime and soda ash are used to remove magnesium noncarbonate hardness.

1. $$MgCl_2 + Ca(OH)_2 \rightarrow \underset{\text{insoluble}}{Mg(OH)_2} + \underset{\text{soluble}}{CaCl_2}$$

This reaction removes the magnesium, but adds dissolved calcium to the water. Soda ash is needed to precipitate it out:

2. $$CaCl_2 + Na_2CO_3 \rightarrow \underset{\text{insoluble}}{CaCO_3} + 2NaCl$$

Often a nominal amount of extra lime is added to the process (referred to as excess lime softening) because stoichiometric calculations and jar testing are usually more efficient than the full-scale process, and a little extra is needed to make the reaction as complete as possible.

Note that this method of softening water will never take all the hardness out. Lime is somewhat soluble in water, and since the process adds calcium to remove calcium, depending upon solubility limits for those removals, there will always be some hardness, along with supersaturated lime and $CaCO_3$, left in the water. It is difficult to bring the total hardness to below 75 mg/l, as $CaCO_3$.

Neither is it desirable to remove all the hardness in municipal supplies. Most utilities prefer to leave the distribution piping with a very slight scale-forming tendency to protect from pipe corrosion, and the leaching of toxic metals from piping and appurtenances. The deposition of a thin layer of scale on the piping depends upon the presence of carbonate alkalinity and some calcium, so that they can combine to form a calcium–carbonate layer.

Reducing the hardness to below 100 to 150 mg/l, as $CaCO_3$, removes most of the nuisance factor. Removal to lower concentrations is not usually considered necessary. This process is costly and difficult to optimize. If noncarbonate hardness is a significantly smaller concentration than carbonate hardness, chemical dosage to remove this part is often omitted.

At completion of the softening operation, the alkalinity should all be in the form of carbonates and hydroxides, and there should be a relatively small amount of each (if settling is optimized). If there is any bicarbonate alkalinity left, not enough lime was added. From an operator's point of view, to remove the calcium the pH must be raised to 9.4. To remove the magnesium the pH must be raised to 10.6.

After softening, the water pH is high and there is supersaturated lime and $CaCO_3$ in the water. It is unstable, often slightly turbid, and excessive deposition of these chemicals will occur in downstream treatment process units and distribution piping. To stabilize the water, CO_2 is added (recarbonation). This acid shifts the alkalinity back to the bicarbonate state, bringing both the remaining hardness and the alkalinity into the stable dissolved form and neutralizing the pH.

$$CaCO_3 + CO_2 + H_2O \rightarrow Ca(HCO_3)_2$$

It is interesting to note that although this recarbonation is an acid addition to the water, it is really a potential alkalinity that is being added. The bicarbonate which results will have buffering power.

Lime is not the only compound useful for softening water supplies. Other hydroxides will provide effective removals by the same method. Magnesium hydroxide has been used, as well as sodium hydroxide. Factors to consider are cost, chemical handling safety, efficiencies achievable, and type of dissolved ions left in the water after process completion.

Analysis: Jar Test for Lime/Soda Ash Dosage for Water Softening

Quality Control

- Maintain water temperature in the jar test the same as in the treatment process, if possible.
- Weigh out the lime needed for each jar separately and add it dry. It is difficult to make a stock solution of lime, which doesn't dissolve, and then add the desired dose into the jars; half of it gets left on the sides of the cylinder or pipet.
- If adding soda ash, prepare a stock solution. Use tap water. The stock should be concentrated enough so that the addition will not substantially increase the volume of water in the jar.
- Allow precipitate to settle a long time (at least 45 min). In the actual treatment process, lime-softening clarifiers are very large and detention times are long.
- Calibrate pH meter with two buffers.

Apparatus

- Jar test apparatus
- pH Meter

Reagents

Lime

CaO

Soda Ash

Na_2CO_3
Weigh out 10 g Na_2CO_3 and dissolve in 1 l tap water.

> NOTE: **Jar testing is a process simulation. Prepare chemicals with water as similar as possible to that used in the treatment process.**

Procedure

Calculate Chemical Dosage:

1. Obtain concentrations of carbon dioxide, total alkalinity, bicarbonate alkalinity, total hardness, and magnesium ion.
2. Change all concentrations to units as $CaCO_3$.
3. Add the concentrations of CO_2, HCO_3^- (or equiv.), Mg^{2+}, and excess lime, if desired.
4. This value is the amount of lime needed, as $CaCO_3$.
5. Convert the concentration back to units as lime. It is lime that is being added, not $CaCO_3$.
6. Adjust for chemical purity.
7. Calculate soda ash dosage based on the concentration of noncarbonate hardness if it is desirable to remove it.

These are the theoretical chemical concentrations needed for softening the water. They are a starting point. By performing the jar test, it can be more accurately determined how well this dosage will soften the water.

8. Set up the jar test apparatus. Fill jars with the water to be softened. Add lime to the jars so that the optimum concentration is added to one of the middle jars. Vary the jar concentrations by 25 to 35 ppm in either direction.
9. If soda ash is desired, dispense optimum concentration to all jars.
10. Set apparatus to rapid mix for 15 sec. Reduce to slow speed for 15 min (just fast enough so that solids don't settle).
11. Turn mixers off. Let the precipitate settle for at least 45 min.
12. Withdraw about 200 ml of sample from each jar from below the surface, taking care not to disturb the precipitate.
13. Test each sample for pH, total hardness, total alkalinity and types of alkalinity.
14. Repeat the jar test procedure, narrowing the concentrations used so that the optimum hardness removal can be more accurately determined.

Federal Limits

None.

Reference

American Water Works Association, *Basic Science Concepts and Applications*, 2nd ed., American Water Works Association, Denver, CO, 1995, p. 442.

Chapter 11

Turbidity

Contents

Turbidity consists of suspended particles in the water and may be caused by a number of materials, organic or inorganic. The occurrence of turbid source waters may be permanent or seasonal. It imparts a cloudy look to the water, and if the particles are colored, may even impart a false color. This is a general designation. It does not specify which type of particles they are, or whether the chemicals the turbidity is composed of are harmful in drinking waters. Substances that commonly cause turbidity in potable water sources are mud, dirt, algae, iron, wastewater solids, and chemical floc.

Treatment Plant Significance

Turbidity has been long known to hinder disinfection by shielding microbes, some of them perhaps pathogens, from the disinfectant. This is the most important significance of turbidity monitoring, and the test for it has been an indication of the effectiveness of filtration of water supplies. Turbidity removal is the principal reason for chemical addition, coagulation, settling, and filtration in potable water treatment. High turbidity waters require more coagulant, create more sludge to handle, and shorten filter runs. They may blind the filter, increase backwashing frequency and the percent of water needed for backwashing. Fine turbidity may bleed through the filter to yield a poor quality finished water and settle solids in the distribution piping.

NOTE: Turbidity shields pathogens from the disinfectant!

Analysis: Turbidimetric Method for Turbidity

The **Jackson Candleometer** was first used to measure turbidity. The calibrated glass tube was slowly filled with water while the chemist looked down through it from above. When the candlelight no longer was visible through the liquid, the calibration at the water surface was read. Turbidity units at that time were called Jackson Turbidity Units (JTU).

Later, the **nephelometer** was developed. A box enclosed a light bulb which directed light at the sample. The amount of light which is scattered at right angles by the turbidity particles was measured, as a measure of the turbidity in the sample, and registered as Nephelometric Turbidity Units (NTU).

The **turbidimeter** is a modern nephelometer. It uses a photoelectric cell to register the scattered light on an analog or digital scale, and the instrument is calibrated with permanent turbidity standards composed of the colloidal substance, formazin. Secondary standards made of a gelled polymer are handy for everyday use, but must be standardized periodically with formazin. Newer versions of this instrument can provide greater accuracy by accounting for natural color in the sample and stray light that escapes the unit.

Quality Control

- Warm up and calibrate instrument with a series of primary standards according to manufacturer's directions. Set values for secondary standards.
- If sample is over 40 NTU, dilute it with distilled water. A turbidimeter is not accurate at high turbidities.
- Let sample warm to room temperature before reading. As it warms, air bubbles will form. Tap them off. They will interfere with the reading.
- Moisture forming on outside of cuvette will interfere. Wipe it off with glass cleaning paper.
- Fingerprints will interfere. Wipe cuvette using glass cleaning paper with a tiny drop of silicone on it. This also minimizes scratches.
- If turbidity reading is not stable, let sample sit a few minutes inside the instrument. Fine particles are still moving around in the sample.
- Settleable solids should not be included in the sample to be measured. True turbidity is composed of permanently suspended particulates. Pour sample into the cuvette **after** particulates have settled.

Apparatus

- Turbidimeter with cuvettes

Reagents

None.

Procedure

1. Measure turbidity of secondary standard. If not accurate according to calibrated value, recalibrate instrument.
2. Pour sample into cuvette; read turbidity.

Federal Limits

0.5 NTU in 95% of samples tested, reported as a monthly average. Reportable readings are taken at the entrance to water distribution system (plant tap) and may never exceed 5 NTU (Surface Water Treatment Rule).

Inline turbidity monitors must be permanently installed at the entrance to the distribution system.

Reference

American Public Health Association, American Water Works Association, and Water Environment Federation, *Standard Methods for the Examination of Water and Wastewater*, 20th ed., American Public Health Association, Washington, D.C., 1998, p. 2-13.

Chapter 12

Chemical Dosage — Jar Testing

Contents

Jar testing, a practical trial-and-error method of determining optimum chemical dosage for turbidity and color removal at water treatment plants, has been the accepted bench testing procedure for many years.

Treatment Plant Significance

Chemical addition at water treatment facilities is performed mainly for the purpose of removing suspended particulate matter from the water. These suspended particles, colloids, are not heavy enough to be settled out and remain stable in suspension unless chemical coagulants are added. Chemical coagulation and flocculation are complicated processes, not always easy to optimize, and even more troublesome if the influent water quality is changeable. Effectiveness depends upon type and amount of coagulant, type, length, and completeness of mixing and flocculation, temperature, pH, alkalinity of the water, other chemicals present, etc.

Some widely used metal coagulants are:

Aluminum Sulfate (Alum): $Al_2(SO_4)_3$

Purchased in two forms: Dry alum $Al_2(SO_4)_3 \cdot 18H_20$ (powder or granular), or liquid alum, available in several strengths.

In Reaction: $Al_2(SO_4)_3 + 3Ca(HCO_3)_2 \rightarrow 2Al(OH)_3 + 3CaSO_4 + 6CO_2$

alum natural alkalinity floc

Alum dosage is optimum within a pH range of 5.5 to 8.5. The chemical itself is acidic and has a tendency to lower the pH of the process water. Adequate natural alkalinity must be present to buffer this added acidity, or an alkaline chemical must be added.

Ferric Chloride: $FeCl_3$

Ferric chloride is available in dry and liquid forms, and the reaction is the same as that for alum; the floc is Ferric Hydroxide $Fe(OH)_3$. Ferric chloride is also very acidic, but works well over a wider pH range than alum does. Its primary disadvantage is its strong yellow color, which stains storage and handling equipment, and presents the possibility of imparting color to the water if dosage is not correct.

Ferric Sulfate: $Fe_2(SO_4)_3$

Ferric sulphate was more widely used years ago than it is today and has been replaced by alum in many treatment plants.

Sodium Carbonate: Na_2CO_3

Soda ash is often used in conjunction with the acidic coagulants to provide the extra alkalinity needed to maintain pH and to protect the distribution piping. It is purchased as a granular solid and has high solubility in water.

Lime: $Ca(OH)_2$

Lime is sometimes used for its settling properties; it acts somewhat differently from the others. This is an alkaline chemical and reacts with the water to form the fine precipitate, $CaCO_3$. This is not a true floc, but has a similar attractive effect on colloidal particles, which group and settle with it. For those waters on which it will work effectively, lime also raises the pH of the water. In its dry form, lime is a very fine powder, is only slightly soluble in water, and must be carried as a slurry. Handling is a nuisance.

Various other metal salts may be used for coagulation. Some sodium and zinc salts will work effectively, but cost limits their use. A disadvantage of all the metal salts used for chemical coagulation is that they are significant additives to the water and increase the amount of necessary sludge handling.

Polymers

Even at the optimum coagulant concentration, destabilization of the colloidal particles is not always complete, and, since the 1950s, addition of a slightly negative polyelectrolyte (polymer) may be favored to complete the neutralization. Polymers are water-soluble, long-chain organic molecules.

There are three basic types of polymers:

Anionic — Negatively charged; used with the metal coagulant for removal of solids in water

Cationic — Positively charged; used for solids/water separation in sludges

Nonionic — No charge; seen more often for industrial uses

Polymers are extremely slippery to the feel. This presents a definite safety hazard when spills occur, either in process or in the laboratory. Most dry polymers have a shelf life of 2 years. If 100% strength liquid, shelf life is 1 year. If diluted, shelf life is about 1 week, though this varies depending upon the amount of dilution. Polymers cannot be effectively dissolved in cold weather. The dry particles cling together and form insoluble "fisheyes." They do not dissolve readily in hot water either, and gel unevenly.

Polymerized Aluminum Compounds

In recent years there has been interest in several newly developed coagulants. Polyaluminum chloride and polyhydroxy aluminum compounds are complexes originating from the initial coagulation that alum begins. Before the aluminum hyroxide particle is formed, these intermediary compounds are captured in colloidal form and can be used as effective coagulants. The coagulant process has already begun when the chemical is added. This effectively shortens reaction time, adds more reactive molecules to the process, and decreases the tendency of the chemical to drop the pH of the water. The process works well at clay/mineral surfaces; the polymers cling readily to turbidity particles, settling them out. Polyaluminum compounds are not effective in every type of water. Their action is hindered in the presence of some organics, particularly fulvic acids and other runoff compounds, but may be a good choice for some.

Analysis: Jar Test Procedure for Chemical Dosage

Jar testing is only an attempt to achieve a ballpark approximation of correct chemical dosage for the treatment process. It is done as a batch operation; both mixing and settling are very efficient. Inplant, however, in a flow-through mode, inefficient mixing and short-circuiting will necessitate adjustments to chemical dosage.

Quality Control

- Keep environmental conditions as close as possible to those in the treatment plant. Sample water temperature in the jars should be kept the same as process water. Laboratory chemical stock solutions should be prepared with tap water or with the type of water that is used in-plant for this purpose.

- Vary dosage of only one coagulant with each jar test run. If polymer is added along with the metal coagulant, hold the polymer dose steady for all jars. Alternately hold metal coagulant dose steady and vary polymer dose.

Apparatus

- Jar test apparatus

Reagents

Coagulant chemical of choice: prepare stock solution of 1000 to 5000 mg/l, dissolved in tap water.

Procedure

1. Set up apparatus with five to six jars. Fill each jar with sample to be tested, leaving enough headspace for mixing. Temperature of sample should be maintained as closely as possible during test.
2. Add increasing amounts of coagulant chemical to each jar so that the range of additions covers the expected optimal chemical dosage.
3. Rapidly mix the jars for about 20 sec.
4. Set mixers to flocculation speed for about 20 min.
5. Stop and withdraw mixers from jars. Allow floc to settle for 30 to 45 min.
6. Decant supernate water from each jar for testing. Be careful not to collect any solids that may be floating on water surface.
7. Test water from each jar for parameter of concern to determine ideal chemical dosage.
8. If necessary, run the jar test procedure again, choosing chemical dosages that narrow the range.

Federal Limits

None.

Reference

American Water Works Association, *Basic Science Concepts and Applications*, 2nd ed., American Water Works Association, Denver, CO, 1995, p. 436.

Chapter 13

Chlorine Residual

Contents

Chlorination is the most widely used means of disinfecting water in this country. When chlorine gas is dissolved into pure water, it forms hypochlorous acid, hypochlorite ion, and hydrogen chloride (hydrochloric acid).

$$Cl_2 + HOH \rightarrow HOCl + (OCl)^- + HCl$$

The total concentration of HOCl and OCl ion is termed **free chlorine residual.** This is the disinfectant. These two exist in equilibrium with each other; if the pH of the water is lower, there will be more HOCl. From a disinfection point of view, this is desirable. HOCl is the more powerful disinfectant of the two.

NOTE: HOCl and OCl^- together are the disinfectant. This is free chlorine residual.

If there is any ammonia in the water, the chlorine will react with the ammonia to form chloramines:

$$Cl_2 + NH_3 \rightarrow NH_2Cl \quad \text{then} \quad NHCl_2 \quad \text{then} \quad NCl_3$$

monochloramine dichloramine trichloramine

With mole ratios of chlorine:ammonia of 1:1, both monochloramine and dichloramine are formed. Concentrations of both are functions of the pH (lower pH, more dichloramine). Trichloramine occurs when greater amounts of chlorine are added, but it is unstable and short-lived. The total concentration of chloramines is termed **combined chlorine residual.** This is not nearly as powerful a disinfectant as free chlorine, and a much larger dose and longer contact time is required to provide the same bactericidal action as free chlorine.

Treatment Plant Significance

Both free and combined chlorine residual are legally acceptable for disinfection at public potable water utilities. The choice is that of the utility, for specific reasons. Free chlorine may react with humic and fulvic acids to form trihalomethanes, carcinogens that are regulated in potable waters. These may be prevented by use of chloramines, by changing point of chlorine application, or by changing to an alternate disinfectant.

Additionally, chloramine disinfection may be preferred if a longer-lasting residual is needed to provide disinfection for an extended distribution system.

Chlorine reacts with many inorganic and organic compounds, oxidizing them. The amount of chlorine needed for these reactions, plus the chlorine spent in disinfection, is called the **chlorine demand**. What is left is the **chlorine residual.**

NOTE: Chlorine Dose = Chlorine Demand + Chlorine Residual

Analysis: DPD Ferrous Titrimetric Method for Chlorine Residual

When N,N-diethyl-*p*-phenylenediamine (DPD) is added to a sample containing free chlorine residual, an instantaneous oxidation of the DPD occurs by the chlorine, producing a DPD compound that has a deep pink color. This color can be titrated out with ferrous ion (ferrous ammonium sulfate, FAS), thereby measuring the quantity of free chlorine residual (ferrous oxidizes to ferric, DPD is reduced back to its original state). A buffer is added to control pH in the 6 to 7 range, at which the reaction works most quickly. If potassium iodide is also added, any chloramines in the sample will oxidize the iodide to free iodine. The iodine will then oxidize the DPD to create more color. Titration with FAS yields a measurement of total chlorine residual. Subtract the results of the two titrations for a measurement of combined chlorine residual.

Quality Control

- Add reagents together **before** adding sample. If not done this way, reagents take 20 min or more to react. It will seem like the test isn't working.
- The chemical reaction to liberate free iodine when measuring the combined chlorine segment is slow. Let sample stand 2 min before titrating.
- This test is accurate for chlorine residual concentrations up to 5 ppm. If sample chlorine concentration is expected to be greater, dilute the sample.
- If color reverts after titration is completed, ignore it.

NOTE: DPD procedures only work for chlorine residual concentrations under 5 mg/l.

Apparatus

- Buret
- Erlenmeyer flask

Reagents

Phosphate buffer solution

pH adjustment

Dissolve 800 mg disodium EDTA in 100 ml distilled water. Add 24 g anhydrous Na_2HPO_4 and 46 g KH_2PO_4. Dilute to 1 l.

DPD indicator solution

Produces the color

Carefully add 6 ml concentrated H_2SO_4 to 2 ml distilled water and mix. Add this to 500 ml distilled water. Add to this 200 mg disodium EDTA and 1 g DPD oxalate. Dilute to 1 l. Store in brown bottle. Discard when discolored.

Ferrous ammonium sulfate

$Fe(NH_4)_2(SO_4)_2 \cdot 6H_2O$

The titrant

Boil 1 l distilled water for 5 min; cool. Bring back to volume. Add 0.3 ml concentrated H_2SO_4 and 1.106 g $Fe(NH_4)_2(SO_4)_2 \cdot 6H_2O$. *Standardization is required.*

Potassium iodide

KI
Converts combined chlorine
Dry crystals

Potassium dichromate

$K_2Cr_2O_7$
To standardize the FAS
Dissolve 0.691 g $K_2Cr_2O_7$ in distilled water and dilute to 1000 ml.

Barium diphenylaminesulfonate

Indicator used to standardize the FAS
Dissolve 0.1 g in 100 ml distilled water.

Sulfuric acid

H_2SO_4
Concentrated

Phosphoric acid

H_3PO_4
Concentrated

Procedure

1. Standardize the FAS titrant:

   Dilute 10 ml concentrated sulfuric acid to 60 ml with distilled water. Add 10 ml of this, 5 ml concentrated phosphoric acid and 2 ml 0.1% barium diphenylamine-sulfonate indicator to a 100-ml sample of FAS and titrate with potassium dichromate to a violet endpoint (100 mg/l is equivalent to 20 ml titrated).
2. Place 5 ml each of buffer and DPD solution in 250-ml Erlenmeyer flask and mix.
3. Add 100 ml sample and mix.
4. For free chlorine residual, titrate rapidly with FAS titrant until color is discharged.
5. For total chlorine residual, omit Step 4. Add several crystals KI and mix. Wait 2 min for reaction to proceed. Titrate to discharge color.
6. 1 ml titrant = 1 mg/l chlorine residual
7. Calculate:

   Total Chlorine Residual – Free Chlorine Residual = Combined Chlorine Residual

Analysis: DPD Colorimetric Method for Chlorine Residual

The pink color produced by the DPD method can be compared to chlorine standards and read with a spectrophotometer according to Beer's law.

Quality Control

- Assure that standards and samples are treated in exactly the same way.

Apparatus

- Spectrophotometer (set at 500 nm)

Reagents

Phosphate buffer solution

pH adjustment
Dissolve 800 mg disodium EDTA in 100 ml distilled water. Add 24 g anhydrous Na_2HPO_4 and 46 g KH_2PO_4 and dilute to 1 l.

DPD indicator solution

Produces the color
Carefully add 6 ml concentrated H_2SO_4 to 2 ml distilled water and mix. Add to 500 ml distilled water. Add to this 200 mg disodium EDTA and 1 g DPD oxalate. Dilute to 1 l. Store in a brown bottle. Discard when discolored.

Potassium iodide

KI
Converts combined chlorine
Dry crystals

Starch

Titration endpoint indicator
Add 5 g starch to 1 l boiling distilled water. Let stand overnight to settle. Decant supernate and use.

Standard sodium thiosulfate

$Na_2S_2O_3 \cdot 5H_2O$
Titrant, to standardize chlorine solution

Dissolve 6.205 g $Na_2S_2O_3 \cdot 5H_2O$ in distilled water. Add 0.4 g solid NaOH and dilute to 1 l. *Must be standardized.*

Potassium bi-iodate

$KH(IO_3)_2$, 0.0021 *M*
To standardize the sodium thiosulfate
Dissolve 812.4 mg $KH(IO_3)_2$ in distilled water and dilute to 1 l.

Sulfuric acid, concentrated

H_2SO_4
Dissolves the tetravalent manganese floc

Chlorine standards

NaOCl
For colorimetric comparison with sample
Dilute commercial chlorine to approximately 100 mg/l.

Acetic acid

CH_3COOH
Concentrated, to standardize chlorine standard

Procedure

1. Standardize the sodium thiosulfate:

   Dissolve 2 g KI in Erlenmeyer flask with 100 ml distilled water.

   Add a few drops concentrated H_2SO_4 and exactly 20 ml potassium bi-iodate solution.

   Dilute to 200 ml and titrate iodine (yellow) with sodium thiosulfate titrant.

   When near the end of titration (pale straw color) add starch (turns blue) and continue titrating endpoint.

   Adjust sodium thiosulfate to 0.025 *M*.

2. Standardize chlorine solution:

   Add 2 ml acetic acid and 25 ml distilled water to a flask.

   Add 1 g KI and 100 ml chlorine solution.

   Titrate with 0.025 *N* $Na_2S_2O_3$ until yellow color is almost gone.

   Add 1 ml starch. Continue titrating until blue color is discharged.

   Run a blank through the titration also.

   Calculate concentration of standard solution:

$$\text{mg/l } Cl_2 = \frac{(\text{ml titrated } Cl_2 \text{ solution} - \text{ml titrated blank}) \times N \text{ sodium thiosulfate} \times 35450}{\text{ml sample}}$$

3. Prepare working standards to cover range of chlorine concentration expected in sample but not greater than 10 mg/l.
4. Set up flasks for each standard, reagent blank, and samples. Add 5 ml DPD and 5 ml buffer to each; mix.
5. Add standards and samples to flasks; mix.
6. If total chlorine residual is desired, also add a few crystals of KI. Wait 2 min for color development.
7. Read absorbance on spectrophotometer. Prepare calibration curve and calculate sample concentration.

Analysis: Amperometric Titration Method for Chlorine Residual

An amperometric titrator consists of a pair of electrodes connected by a salt bridge. One electrode senses the concentration of ionized chlorine in the solution and becomes polarized by it. A current flow then occurs to the opposite electrode, which is picked up by a microammeter, producing a needle reading. The intensity of this reading decreases steadily (needle drops) as the ionized chlorine is neutralized by the reagent, phenylarsine oxide. When needle movement stops the current has stopped, and the chlorine has all been neutralized. At that point the titration is ended, and ml used is recorded. The change in amperage is observed as free chlorine is reduced by the titrant phenylarsine oxide. The reaction is sluggish over pH 7.5, and a pH 7 buffer is added to control pH.

At pH levels below 6, chloramines are reduced and can be measured indirectly (iodometrically). If KI is added, the chloramines oxidize iodide to free iodine. The phenylarsine oxide titrant reduces the free iodine, thereby measuring the amount of chloramines present. The pH is controlled for this titration with a pH 4 buffer.

By conducting a two-stage titration, first at pH 7, then with KI at pH 4, free and combined chlorine can be separated.

Quality Control

- Amperometric titration requires a little practice in endpoint determination. With new samples, repeat procedure at least once.
- This method measures accurately down to 0.01 mg/l chlorine residual and is not subject to interference from color or turbidity.
- Amperometric titrator electrodes may need periodic cleaning with an abrasive non-chlorine cleanser. See manual provided with electrodes.
- Buffers used in this procedure must be recently prepared or purchased. If bacterial growth has occurred in the buffers, a chlorine demand will be established that will lead to inaccurate results.

Apparatus

- Amperometric titrator

Reagents

Phenylarsine oxide

0.00564 *N*
The titrant
This reagent is best purchased from a chemical distributor as a solution. Preparation is tedious, and safe handling of the highly toxic powder requires extreme care.

pH 4 buffer

To titrate combined chlorine residual

pH 7 buffer

To titrate free chlorine residual

Potassium iodide solution

KI
Converts to free iodine for combined chlorine titration
Dissolve 50 g KI in distilled water. Dilute to 1 l. Store in a dark bottle.

Procedure

1. **For free chlorine residual:** Add 1 ml pH 7 buffer to 200-ml sample. Titrate with phenylarsine oxide in progressively decreasing volumes, observing current change on titrator. Stop titrating when current stops changing. Record ml titrated.
2. **For combined chlorine residual:** To remaining sample add 1 ml KI and 1 ml pH 4 buffer. Continue titrating to endpoint, as above.
3. Total ml titrated = total chlorine residual

   Total ml titrated – ml titrated #1 = combined chlorine residual

NOTE: Phenylarsine oxide is toxic. Handle carefully.

Analysis: Ion-Specific Electrode Method for Chlorine Residual

Measurement by ion-specific electrode is an indirect (iodometric) measurement of chlorine. The electrode develops a voltage that is dependent upon the difference in concentrations of iodine and iodide in the solution. KI and acid buffer are added to the sample. Chlorine in the sample changes the KI to free iodine. KIO_3 is used as a standard instead of chlorine; it is converted to iodine, which is measured. This method can only be used to determine total chlorine residual.

Quality Control

- Add reagents and wait 2 min.
- Mix standard, acid, and KI together first. This shortens the waiting time.
- The stock KI solution should be colorless. If it turns yellow at all, discard.

Apparatus

- pH/millivolt meter
- Combination reference/iodine selective electrode (for chlorine residual)

Reagents

pH 4 buffer

Potassium iodide solution

KI
Converts to free iodine for combined chlorine detection
Dissolve 42 g KI and 0.2 g Na_2CO_3 in 500 ml distilled water. Store in a dark bottle.

Potassium iodate solution

KIO_3, 0.00281 *N*
The standard
Dissolve 0.1002 g KIO_3 in distilled water and dilute to 1 l.

Procedure

1. Pipet 0.2, 1.0, and 5.0 ml potassium iodate solution (0.2 mg/l, 1.0 mg/l, and 5.0 mg/l) into 100-ml stoppered volumetric flasks.
2. Add to each flask 1 ml pH 4 buffer and 1 ml KI solution. Mix. Let stand 2 min.
3. Dilute each to 100 ml with distilled water; mix. Pour into beakers and insert electrode into first standard; record millivolts.
4. Rinse electrode and repeat reading for samples, using 100 ml.
5. Prepare calibration curve on semilog graph paper.

Federal Limits

A detectable chlorine residual must be present throughout the water distribution system.

References

American Public Health Association, American Water Works Association, and Water Environment Federation, *Standard Methods for the Examination of Water and Wastewater*, 20th ed., American Public Health Association, Washington, D.C., 1998, p. 4-53.

American Water Works Association, *Basic Science Concepts and Applications*, 2nd ed., American Water Works Association, Denver, CO, 1995, p. 478.

Chapter 14

Chlorine Demand

Contents

The chlorine demand of a water is the difference between the amount of chlorine applied and the amount of free, combined, or total chlorine residual remaining at the end of the contact period.

$$\text{Dose} = \text{Demand} + \text{Residual}$$

Chlorine demand varies with dose, residual, contact time, pH, and temperature. Organic and inorganic chemicals react with chlorine, which oxidizes or combines with them. Chlorine also disinfects. The total amount of chlorine used up in all these reactions is the chlorine demand.

Treatment Plant Significance

Determination of chlorine demand is an important design consideration. It is used to determine number and capacity of chlorinators required, amount of chlorine needed, type of containers, and all appurtenances required for handling and storage. It enables calculation of chlorine dose to ensure effective disinfection without waste of chlorine.

Chlorine residual, with time, will dissipate, because disinfection continues after the water has entered the distribution system. When no residual is desired, as in prechlorination for taste and odor control at potable water treatment plants, the demand and the dose are equal.

Analysis: Jar Test Procedure for Chlorine Demand

A series of jars are set up, each with the same amount of sample. Increasing dosages of OCl ion are added to the jars. A contact time is allowed, similar to what is designed into the treatment plant. Chlorine residual is then measured in each jar and demand is calculated.

If ideal dosages have been chosen, the first of the jars should have no residual. The last of the jars should have a significantly larger residual than desired. Among the middle jars would be the one with the dose which provides the desired residual at the correct contact time.

Quality Control

- Dilute the stock solution with distilled water if chlorine gas is used in plant. There should be no chlorine demand or active chlorine in the dilution water.
- If sodium or calcium hypochlorite are being used in plant, dilute the stock solution with the type of water that it is diluted with in the plant.
- Stagger the chlorine additions to each jar so that there will be time to test each chlorine residual, leaving exactly the same amount of contact time for each.
- Keep the sample temperatures as close as possible to the water temperature in plant.

Apparatus

- Jar test apparatus

Reagents

Chlorine stock solution

NaOCl
For chlorine addition to sample
Dilute commercial chlorine to approximately 100 mg/l. *Must be standardized.*

Acetic acid

CH_3COOH
Concentrated, to standardize chlorine standard

Potassium iodide

KI
Dry crystals, converts combined chlorine

Starch

Titration endpoint indicator
Add 5 g starch to 1 l boiling distilled water. Let stand overnight to settle. Decant supernate and use.

Standard sodium thiosulfate

$Na_2S_2O_3 \cdot 5H_2O$
Titrant, to standardize chlorine solution
Dissolve 6.205 g $Na_2S_2O_3 \cdot 5H_2O$ in distilled water. Add 0.4 g solid NaOH and dilute to 1 l. *Must be standardized.*

Potassium bi-iodate

$KH(IO_3)_2$, 0.0021 *M*
To standardize the sodium thiosulfate
Dissolve 812.4 mg $KH(IO_3)_2$ in distilled water and dilute to 1 l.

Sulfuric acid, concentrated

H_2SO_4
Dissolves the tetravalent manganese floc

Procedure

1. Standardize the sodium thiosulfate:

   Dissolve 2 g KI in an Erlenmeyer flask with 100 ml distilled water.

   Add a few drops concentrated H_2SO_4 and exactly 20 ml potassium bi-iodate solution.

   Dilute to 200 ml and titrate iodine (yellow) with sodium thiosulfate titrant.

   When near the end of titration (pale straw color) add starch (turns blue) and continue titrating endpoint.

   Adjust sodium thiosulfate to 0.025 *M*.

2. Standardize chlorine solution:

   Add 2 ml acetic acid and 25 ml distilled water to a flask.

   Add 1 g KI and 100 ml chlorine solution.

   Titrate with 0.025 *N* $Na_2S_2O_3$ until yellow color is almost gone.

   Add 1 ml starch. Continue titrating until blue color is discharged.

   Run a blank through the titration also.

   Calculate concentration of standard solution:

$$\text{mg/l } Cl_2 = \frac{(\text{ml titrated } Cl_2 \text{ solution} - \text{ml titrated blank}) \times N \text{ sodium thiosulfate} \times 35450}{\text{ml sample}}$$

3. Dispense the same amount of sample into each jar. Add increasing amounts of chlorine to each, starting with a concentration which will not satisfy the demand and will provide no residual, and ending with a concentration which should provide a greater than desired residual.
4. Mix rapidly for a few seconds. Stop stirrers and wait until allotted contact time has elapsed.
5. Perform chlorine residual analysis on all jars. Calculate chlorine demand for each jar which showed a chlorine residual. See Chapter 13.

> NOTE: **The chlorine demand should come out the same in all jars for which it is calculated, regardless of what the residual is. It has to. It's the same sample.**

Federal Limits

None.

References

American Public Health Association, American Water Works Association, and Water Environment Federation, *Standard Methods for the Examination of Water and Wastewater*, 20th ed., American Public Health Association, Washington, D.C., 1998, p. 2-40.

American Water Works Association, *Basic Science Concepts and Applications*, 2nd ed., American Water Works Association, Denver, CO, 1995, p. 478.

Chapter 15

Chloride

Contents

Chlorides occur in all natural waters in varying concentrations. Concentration is usually greater in groundwater than surface water, especially if salt deposits are in the area. Fresh water which has any contact with sea water or sea air will be high in chloride.

Chlorides in small concentrations are not harmful to humans in drinking water, and with some adaptation the human body can tolerate a water with as much as 2000 mg/l chloride ion. However, above a concentration of 250 mg/l chloride, the water may taste salty, and a secondary MCL has been set at this level for community supplies. Since humans generally consume more salt than the body needs, the excess is excreted, and domestic wastewaters usually range 20 to 80 mg/l higher in chloride than the carriage water.

Treatment Plant Significance

An increase in chloride concentration in a drinking water source may indicate sewage contamination or an industrial waste discharge. Care must be taken in interpreting this, however, to distinguish it from the natural chloride concentration. Overpumping water wells near the ocean may cause saltwater intrusion, yielding salty water. Chloride is an excellent electrical conductor. High chloride content in process waters may promote pipe corrosion.

Removal of chloride from potable and industrial process waters is very difficult, and generally requires desalination. An ion exchange medium has also been developed which will remove chloride, but it is expensive and not widely used.

Analysis: Argentometric Method for Chloride

This is a titrimetric method. An indicator, potassium chromate, is added to the sample, which loosely binds up a few chloride ions. Silver nitrate solution is then titrated into the sample, producing the fine white precipitate, silver chloride. When all the free chloride ions are complexed in this reaction, the silver nitrate takes the bound chloride ions from the indicator, producing the blood red precipitate, silver chromate, marking the end of the titration. Diagrammatically, the reaction looks like this:

$$AgNO_3 + Cl^- \xrightarrow{K_2CrO_4 \text{ indicator}} \underset{\text{white}}{AgCl} + \underset{\text{red}}{Ag_2CrO_4}$$

A standard of 500 mg/l chloride ion is run along with the sample analysis and standardized as part of the procedure.

Quality Control

- Standardize titrant.
- Titrate carefully using white background. When the red Ag_2CrO_4 is visible throughout the solution, end the titration.

Apparatus

- Buret

Reagents

Potassium chromate

K_2CrO_4
The indicator

Dissolve 50 g K_2CrO_4 in a little distilled water. Add silver nitrate solution a drop at a time until a definite red precipitate is formed. Let stand 12 h, filter, and dilute to 1 l.

Standard silver nitrate

$AgNO_3$, 0.0141 *M*
The titrant
Dissolve 2.395 $AgNO_3$ in distilled water and dilute to 1000 ml
Standardize against NaCl.

Standard sodium chloride

NaCl, 0.0141 *M*
Dissolve 824 mg NaCl (dried at 140°C) in distilled water and dilute to 1000 ml. Concentration of this solution is 500 mg/l.

NOTE: Silver nitrate stains skin. Avoid contact.

Procedure

1. Measure 100 ml sample into an Erlenmeyer flask. Treat the NaCl standard solution as if it were a sample, and run this through the procedure also. The solution will turn milky white and then pale pink. End titration at this point. Also, run a distilled water blank.
2. Add 1 ml K_2CrO_4 indicator; mix.
3. Titrate with $AgNO_3$ titrant until red precipitate diffuses throughout the solution. End titration and record ml used.
4. Calculate:

$$\text{mg Cl}^-/\text{l} = \frac{(A - B) \times N \times 35450}{\text{ml sample}}$$

A = ml titrated for sample
B = ml titrated for blank
N = normality of titrant

Calculate first for the titrant normality using ml titrated for the NaCl standard; mg Cl^-/l is 500. Calculate for *N*.

When calculating mg Cl^-/l for the samples, use the titrant normality calculated above.

NOTE: The silver chloride precipitate formed in a water sample of low chloride content is barely visible. It may not be noticed. However, it will be highly visible in the 500 mg/l standard, and make the solution look milky. Titrate until the milky solution turns pink.

Analysis: Ion-Specific Electrode Method for Chloride

This potentiometric method is useful in colored or turbid waters where a titration endpoint based on a color change may be hard to detect.

Quality Control

- Check electrode accuracy according to manufacturer's directions for millivolt change per decade of concentration.
- This method requires some skill to acquire an accurate endpoint. Several trials may be necessary.

Apparatus

- pH meter that reads in millivolts
- Chloride electrode + pH reference electrode; alternatively, use combination electrode
- Stirring mechanism

Reagents

Standard silver nitrate

$AgNO_3$, 0.014 *M*
The titrant
Dissolve 2.395 $AgNO_3$ in distilled water and dilute to 1000 ml. *Standardize against NaCl.*

Standard sodium chloride

NaCl, 0.0141 *M*
Dissolve 824 mg NaCl (dried at 140°C) in distilled water and dilute to 1000 ml. Concentration of this solution is 500 mg/l.

Nitric acid

HNO_3
Concentrated

Procedure

1. Set instrument to millivolt mode.
2. Standardize titrant: Place 10 ml standard NaCl in a 250-ml beaker and dilute to 100 ml. Add 2 ml concentrated HNO_3. Immerse stirrer and electrodes; start stirrer. Add standard $AgNO_3$ titrant in increments, recording millivolts with each addition. Start with large increments, then smaller as the titration progresses. Determine volume of titrant used at the point which shows the greatest millivolt reading. Record this volume (ml used).

3. For sample analysis, use 100-ml sample (or a portion containing not more than 10 mg/l chloride). Add concentrated nitric acid a drop at a time until acid to litmus paper. Add 2 ml more. Immerse electrodes; start stirrer. Add standard $AgNO_3$ titrant in increments, recording millivolts with each addition. Start with large increments, then smaller as the titration progresses. Determine volume of titrant used at the point that shows the greatest millivolt reading. Record this volume (ml used).
4. For the most accurate work, run a distilled water blank through the procedure also.
5. Calculate:

$$\text{mg Cl}^-/\text{l} = \frac{(\text{A} - \text{B}) \times N \times 35450}{\text{ml sample}}$$

A = ml titrated for sample

B = ml titrated for blank

N = normality of titrant

Calculate first for the titrant normality using ml titrated for the NaCl standard; mg Cl^-/l is 500. Calculate for N.

When calculating mg Cl^-/l for the samples, use the titrant normality calculated above.

Federal Limits

250 mg/l (secondary MCL).

Reference

American Public Health Association, American Water Works Association, and Water Environment Federation, *Standard Methods for the Examination of Water and Wastewater*, 20th ed., American Public Health Association, Washington, D.C., 1998, p. 4-66.

Chapter 16

Specific Conductance

Contents

Specific conductance (conductivity) is a measurement of the capacity of water to carry an electric current. It will vary according to the amount of ionized substances in the water, and the temperature at which the measurement is made.

Treatment Plant Significance

This test is not routine in potable water treatment, but when performed on source water is a good indicator of contamination. Although it measures primarily for inorganic substances, the concentration of these, along with the organic pollutants, will be elevated if contamination exists. Conductivity readings can be used to indicate wastewater contamination or saltwater intrusion. Distilled water used for potable water analyses at public water supply facilities must have a conductivity of no more than 1 μmho/cm. This is an EPA regulation applying to certification of potable water laboratories.

Analysis: Instrumental Method for Specific Conductance

The conductivity meter is a resistance meter, measuring ohms. The electrode (conductivity cell) is electroplated with platinum to improve contact with the solution, and cells with a range of sensitivities can be chosen. Cells with small electrodes spaced further apart are used for waters with very high conductivities, i.e., hard waters and wastewaters. The meter actually measures resistance, converts it to conductance (reciprocal), and then multiplies it by a constant to convert it to a conductivity reading in μmhos.

Conductivity is measured in units called micromhos (μmhos/cm) or micro-Siemens/cm (μS/cm). The numerical equivalence between these two is 1:1.

Typical readings are:	Distilled water:	1 to 5 μmhos
	Raw and tap water:	50 to 500 μmhos
	Hard waters:	500 to 2000 μmhos

> NOTE: **Micromhos can be equated to units of mg/l, as NaCl, on approximately a 2:1 basis: 2 μmhos = 1 mg/l, as NaCl.**

Quality Control

- Temperature is very important in this test. Conductivity can vary by as much as 10 μmhos for every 1°C change. It is best to bring the sample to room temperature before measuring.
- Calibrate meter with purchased NaCl conductivity standards.

Apparatus

- Conductivity meter with electrode appropriate for conductivity range to be measured

Reagents

Purchased conductivity standards

NaCl

Procedure

1. Rinse cell well in distilled water.
2. Immerse cell in conductivity standards to check calibration. Adjust if necessary.
3. Immerse cell in sample. Read conductivity.

Federal Limits

None.

Reference

American Public Health Association, American Water Works Association, and Water Environment Federation, *Standard Methods for the Examination of Water and Wastewater*, 20th ed., American Public Health Association, Washington, D.C., 1998, p. 2-44.

Chapter 17

Langelier Index

Contents

Variations of this test has been performed since ancient times. Over the centuries it has taken many names:

Langelier Index
$CaCO_3$ Saturation Index
Ryzner's Index
Corrosivity Index
Stability Index
The Marble Test

It refers most directly to measurement of the scale forming tendency of a water. The influence of hardness, alkalinity, pH, and many other components of the water come into play here. This test is meant only to give the utility personnel an approximation of whether the water has a tendency to be corrosive or scale forming.

Treatment Plant Significance

The significance of the Langelier Index is applied to the water distribution system and the condition of the piping interior. Most water treatment plants would like to obtain a *very slight* scale-forming tendency in the water — just enough to put a fine coating of calcium carbonate on interior surfaces. Operations personnel adjust chemical feed rates to achieve this condition. Treatment plants which add soda ash along with their metal coagulant are adding it partly for this reason — to protect the piping.

Lime/soda ash softening leaves the water in an unstable and scale-forming condition. Recarbonation is needed to stabilize the water. Ion exchange softening removes all the calcium and magnesium which could provide a slight scale-forming tendency. This water is considered corrosive. If this method of softening is used, split treatment is often employed to minimize the effect.

Careful control is needed. Too great a scale-forming tendency will reduce effective diameter of piping, generating greater head loss, and allowing less flow. The condition of interior surfaces within the distribution system is a very important consideration to a water utility. Water quality will affect all pipes and appurtenances of the system.

Analysis: pH Method for Langelier Index

If a water is corrosive, addition of $CaCO_3$ would, in neutralizing the acids, be shifted to the soluble bicarbonate state. Any left after acids are neutralized would precipitate out, since $CaCO_3$ is insoluble. Since acids had been neutralized, the pH would rise. Subtraction of pH at the saturation point of $CaCO_3$ from original pH of the water would yield a negative number.

If a water is scale-forming, $CaCO_3$ is already supersaturated in the water, present in an unstable form, with the tendency to slowly precipitate out. Addition of more $CaCO_3$ to the sample will shock it, taking out even more $CaCO_3$, and leaving the sample in a more stable state with a lower pH. Subtraction of pH at the saturation point of $CaCO_3$ from original pH would yield a positive number.

Quality Control

- Calibrate pH meter with two buffers

Apparatus

- pH meter

Reagents

Calcium carbonate

$CaCO_3$

Procedure

1. Put a sample of water in an Erlenmeyer flask. Take the pH of the water.
2. Add a small scoop of $CaCO_3$; mix and let settle. Take the pH again.
3. Calculate:

$$\mathbf{pH - pH_s = Langelier\ Index}$$

If the result is a negative number, the water is corrosive.

If it is positive, the water has a scale-forming tendency. Results that are less than one whole number mean that the tendency is very slight.

A result of 0 means that the water is neither corrosive nor scale forming.

Analysis: Total Alkalinity Method for Langelier Index

With corrosive waters, addition of $CaCO_3$ would, in neutralizing the acids, be shifted to the soluble bicarbonate state. Any left after acids are neutralized would precipitate out, since $CaCO_3$ is insoluble. Since acids had been neutralized, the total alkalinity would rise. Subtraction of alkalinity at the saturation point of $CaCO_3$ from original alkalinity would yield a negative number.

With scale-forming waters, $CaCO_3$ is already supersaturated in the water, present in an unstable form, with the tendency to slowly precipitate out. Addition of more $CaCO_3$ to the sample will shock it, taking out even more $CaCO_3$, and leaving the sample in a more stable state with a lower total alkalinity. Subtraction of alkalinity at the saturation point of $CaCO_3$ from original alkalinity would yield a positive number.

Quality Control

- Standardize acid titrant carefully.
- Titrate carefully. Don't overrun the endpoint.
- If using pH meter, warm it up for about 30 min before using.
- Calibrate meter with two pH buffers.

Apparatus

- Buret
- pH meter

Reagents

Calcium carbonate

$CaCO_3$

Sulfuric acid

H_2SO_4, 0.02 *N*
The titrant
Dissolve 1.4 ml concentrated sulfuric acid in 1 l distilled water. This is approximately 0.05 *N* acid. *Must be standardized.*

Sodium carbonate

Na_2SO_4, 0.05 *N*
Used to standardize the acid
Dry 3 to 5 g Na_2CO_3 at 250°C for 4 h and cool in a desiccator. Weigh 2.5 g dissolved in distilled water and fill to 1 l.

Methyl orange

The 4.5 pH indicator
Dissolve 500 mg methyl orange powder in distilled water and dilute to 1 l. Color change is: Orange → Rose at pH 4.5.

Bromocresol green

Alternate 4.5 pH indicator
Dissolve 100 mg bromocresol green, sodium salt, in 100 ml distilled water. Color change is: Blue → Yellow Green at pH 4.5.

Procedure

1. Standardize sulfuric acid:

   This procedure allows determination of the exact normality of the prepared sulfuric acid, and to adjust it to exactly 0.02 *N*.

   With 50 ml Na_2CO_3 in a small Erlenmeyer flask, titrate the sulfuric acid into it, using the pH meter. If the acid is exactly 0.05 *N*, then 50 ml of the titrant should bring the Na_2CO_3 solution to the pH 4.5 inflection point (neutralized).

   However, both solutions were dissolved in distilled water that has CO_2 (an acid) in it. To eliminate the CO_2, proceed with the titration down to pH 5. Then boil the solution for 5 min under a watch glass. Cool the solution (ph should go up a little).

   Continue the titration, down to the inflection point; record total ml acid used.

   Calculate: $C \times V = C \times V$ for the true normality of the acid.

   acid carbonate

   Dilute the acid down to 0.02 *N* for use in the total alkalinity test.

2. Measure 100-ml sample in a 250-ml Erlenmeyer flask. Add a few drops of methyl orange or bromocresol green indicator and mix.

3. Titrate 0.02 *N* sulfuric acid into sample, swirling flask until endpoint color change is noted. Record ml titrated.
4. Calculate: mg/l total alkalinity as $CaCO_3$ = $\frac{\text{ml titrated} \times 1000 \text{ m/l}}{\text{ml sample}}$
5. Put a sample of water in an Erlenmeyer flask. Take the total alkalinity of the water.
6. Add a small scoop of $CaCO_3$. Mix and let settle. Take the total alkalinity again.
7. Calculate:

$$\frac{(\textbf{Tot. Alk} - \textbf{Tot. Alk}_s) \times \textbf{1000}}{\textbf{ml sample}} = \textbf{Langelier Index}$$

If the result is a negative number, the water is corrosive.

If it is positive, the water has a scale-forming tendency.

Results that are less than one whole number mean that the tendency is very slight.

A result of 0 means that the water is neither corrosive nor scale forming.

Federal Limits

0.0 (secondary MCL).

References

American Public Health Association, American Water Works Association, and Water Environment Federation, *Standard Methods for the Examination of Water and Wastewater*, 20th ed., American Public Health Association, Washington, D.C., 1998, p. 2-29.

American Water Works Association, *Basic Science Concepts and Applications*, 2nd ed., American Water Works Association, Denver, CO, 1995, p. 452.

Chapter 18

Color

Contents

Color in water most often originates from organic sources: decomposition of forest debris, such as leaves, evergreen needles, etc. Organic compounds such as tannins and lignins dissolve into the water. Some organics bond to iron to produce soluble color compounds. Decomposing algae from a recent bloom may cause significant color. Pollution from highly colored wastewater, treated or untreated, may color a natural water body. Possible inorganic sources of color are salts of iron (yellow), copper (blue), and potassium permanganate (pink) added in excess at the treatment plant. It is less likely that the color of the water will be from these sources than from organic sources.

True color is dissolved. It is measured colorimetrically and compared against an EPA color standard.

Apparent color may be caused by suspended material (turbidity) in the water. Though it may also be objectionable in the water supply, it is not meant to be measured in the color test. Particulate oxidized iron is probably the most common cause of apparent color. The term "red water" refers to this condition.

Treatment Plant Significance

Color of itself has no health significance in drinking waters. A secondary MCL is set at 15 color units, and it is recommended that community supplies provide a water that has less color. Approximately 20 color units is visually noticeable. Industrial process water may need treatment to remove natural color, depending upon the industry. Color removal at the treatment plant, if necessary, will depend on the source of the color.

Alum or ferric chloride coagulation is often effective. It removes apparent color (settles out the turbidity) and often much of the true color. Some dissolved colored compounds are converted to particulates and then settled out. Some are converted to colorless dissolved forms. Lowering the pH of the water may decrease the color. Oxidation of color causing compounds to a noncolored version is sometimes effective. Activated carbon treatment may adsorb some of the organics causing color. If the problem is with apparent color, filtration should trap the colored particles.

NOTE: True color is dissolved. It is not caused by particles in the water.

Analysis: Colorimetric Method for Color

It is best to filter the sample first to remove any apparent color. Purchased Standard Color Solution composed of potassium chloroplatinate (K_2PtCl_{16}) tinted with cobalt is used for comparison. This has a color similar to most natural color. The unit labels, **color units** is used (equivalent to 1 mg/l of the standard, as K). No reagent is needed for the color analysis. It is only the water color that is being read on the spectrophotometer.

Readings are taken using the transmittance mode. Because there is usually very little color in water, a larger positive number will result if transmittance is used.

Quality Control

- Filter the sample using a 0.45-µm membrane filter to eliminate apparent color.

Apparatus

- Spectrophotometer set at 320 nm

Reagents

Standard Color Solution

K_2PtCl_{16} (500 color units)

Procedure

1. Set the instrument to 100% transmittance using distilled water.
2. Filter the sample with a 0.45-µm membrane filter.
3. Dilute purchased color standard to a range of working standards expected to cover sample color concentration.
4. Prepare graph with color units across the bottom (X axis), and 100% transmittance at the bottom of the Y axis. To calculate slope, measure the **distance** from the origin (100% transmittance) to the plotted point on the Y axis. To calculate sample concentration (Conc. = Transmittance × Slope Number), again measure the distance from the origin to the transmittance point of the sample.

Federal Limits

15 color units (secondary MCL).

Reference

American Public Health Association, American Water Works Association, and Water Environment Federation, *Standard Methods for the Examination of Water and Wastewater*, 20th ed., American Public Health Association, Washington, D.C., 1998, p. 2-1.

Chapter 19

Odor

Contents

Odor in drinking water is a general designation and may be caused by a number of constituents. Ideally, water should be odorless, but algae (especially diatoms and blue green algae), actinomycetes bacteria, phenols, sulfur compounds, and ammonia are common causes of odor in waters, either seasonally or continually. There are a number of volatile organic contaminants that may impart an odor; some may be toxic. Hazard to health will depend on the chemical properties of the contaminant. Odor has no significance to health and is merely a nuisance factor, but is objectionable in drinking water. Odor is very related to taste. Most of the "taste" we sense is due to the odor. If the source is organic in nature, there is often a color associated as well.

Treatment Plant Significance

Removal depends upon the source of the odor. Some organic substances that cause odor can be removed with powdered activated carbon. If the odor is caused by a gas, scrubbing (aeration) may remove it. Some odor-causing chemicals can be oxidized to odorless chemicals with chlorine, potassium permanganate, or other oxidizers.

> **NOTE: Chlorine can intensify certain odors: some algae, actinomycetes, and phenols. Chlorine itself, especially in combination with ammonia, may have an odor.**

Coagulation and settling may remove some particulates which, when later dissolved in the water, may have potential odor-causing capacity. In distribution, odors may be generated at dead ends and slow-moving areas.

Analysis: Determination of Threshold Odor Number

The test for odor has no scientific means of measurement and is not very accurate. Odor will dissipate upon standing, even in a closed bottle. Refrigeration of the sample will minimize odors; heating usually exaggerates it. If the odor potential is caused by a gas or volatile organic as the sample is heated, odor compounds will be emitted. In a short time, however, they will have dissipated, leaving the sample fairly odorless. It is easy to miss it, even during the test.

At potable water treatment facilities a sample is routinely heated to 60°C. Odor is observed and recorded. At intervals, a threshold odor number (TON) may be required.

Quality Control

- Odor-free water must be prepared for comparison. Check its quality by running it through the odor test.
- Reserve special flasks for odor testing alone. Don't perform other test procedures in these flasks.
- As samples in flasks heat, air inside expands. Don't let the top pop off or the odor will be lost.
- Determine odor as soon as possible after warming flasks. If they stand on the counter cooling, a vacuum will form and pull at the tops, making it difficult to remove them.
- When sniffing odor flasks, do it quickly and record your first impression. Sniffing repeatedly will cause confusion.

Apparatus

- Six 500-ml Erlenmeyer flasks with ground glass tops

Reagents

Odor-free water

For sample comparison

Activated carbon

To prepare odor-free water

Procedure

1. Prepare odor-free water: Pass distilled water through a column of activated carbon at a rate of 100 ml/min.
2. Line up the six flasks. Mark one as a blank and fill with 200 ml odor-free distilled water. Mark the others 1 to 5 with small numbers (so that they are not noticeable).
3. Dispense 200-ml volumes into each of the five flasks:
   - #1: 200-ml sample
   - #2: 100-ml sample, 100 ml distilled water.
   - #3: 50-ml sample, 150 ml distilled water.
   - #4: 20-ml sample, 180 ml distilled water.
   - #5: 5-ml sample, 195 ml distilled water.
4. Close all flasks. Heat to 60°C in water bath. Use the blank to check the water temperature with a thermometer. Don't open the others.
5. Remove all flasks from bath and place on counter. Shuffle flasks so that the numbers are mixed up. Don't look at the numbers. Open each flask and sniff. Compare to blank.
6. Record each number and whether or not an odor was noticed in that flask.
7. Repeat the entire procedure using a range of dilutions in which the flask with the least odor is the one with the greatest amount of sample. Dilute down from this.
8. Calculate:

$$\text{TON} = \frac{\textbf{Total Volume of Water Sample}}{\textbf{Lowest Sample Volume with Odor}}$$

> **NOTE: Each person has a different capacity for noticing odor. Those with colds or any kind of nasal blockage, those who smoke, those who work or play in high odor areas, are not likely to be very effective odor testers.**

Descriptions of Odors

Code	Nature of Odor	Possible Source
Ac	Aromatic/cucumber	Synura
Bg	Balsam/geranium	Asterionella
Bn	Balsam/nasturtium	Aphanizomenon
Cc	Chemical/chlorinous	Chlorine
Ch	Chemical/hydrocarbon	Oil refinery wastes
Cm	Chemical/medicinal	Phenol
Cs	Chemical/sulfur	Hydrogen sulfide
Df	Disagreeable/fishy	Uroglenopsis, Dinobryon
Dp	Disagreeable/pigpen	Anabena
Ds	Disagreeable/septic	Stale sewage
Ep	Earthy/peaty	Peat
G	Grassy	Crushed grass
M	Musty	Decomposing straw
Mm	Musty/moldy	Actinomycetes

Federal Limits

3 TON (secondary MCL).

Reference

American Public Health Association, American Water Works Association, and Water Environment Federation, *Standard Methods for the Examination of Water and Wastewater*, 20th ed., American Public Health Association, Washington, D.C., 1998, 2-12.

Chapter 20

Fluoride

Contents

Fluorine is a very reactive gas that is not found in the free state naturally. Of volcanic origin, it reacts with the rocks and its reactions go to completion, forming stable fluoride salts. Fluorides are used to make glass and ceramics. Since 1940, fluoride has been added to many community water supplies throughout the United States to prevent dental caries and to aid children's growing teeth. Grand Rapids, Michigan is believed to have been the first community to add fluoride to its water supply.

Tooth enamel is composed of a mixture of calcium carbonate and calcium hydroxide. Fluoride reacts with the calcium in teeth to produce calcium fluoride, a substance that is harder than either of these two.

Treatment Plant Significance

Low fluoride waters: The chemicals used in treatment process for fluoride addition are:

NaF:	Sodium fluoride, solid
Na_2SiF_6:	Sodium silicofluoride, solid
H_2SiF_6:	Hyrofluosilicic acid; most widely used; purchased as a concentrated acid

High fluoride waters: Some natural waters are high in fluoride, particularly in the Southwest where volcanic deposits may occur. If these waters are to be used as public water supplies, the fluoride must be removed. Activated alumina has been used to remove fluoride. Ion exchange processes can be effective, and desalination (reverse osmosis) removes all ions. All of these treatments are costly.

Drinking waters with fluoride concentration over 2 mg/l may cause darkening of teeth (dental fluorosis). Drinking waters with fluoride over 4 mg/l may produce skeletal damage. A concentration of 1 to 1.5 mg/l fluoride is optimum.

Analysis: Colorimetric Method for Fluoride

Fluoride ion reacts with zirconium ion and produces zirconium fluoride, which bleaches an organic red dye, SPADNS, in direct proportion to its concentration. This can be compared to standards and read colorimetrically.

> NOTE: **The more concentrated the fluoride, the *lighter* the color becomes (SPADNS method).**

Quality Control

- Aluminum, chlorine, and polyphosphates may cause interference with this method. Sample should be distilled first in treatment plants using alum as a coagulant.
- Fluoride etches glass. Store concentrated solutions in plastic bottles.

Apparatus

- Spectrophotometer, 570 nm

Reagents

Sodium fluoride

NaF, 10 mg/l
The standard
Dissolve 221 mg anhydrous sodium fluoride in distilled water and dilute to 1000 ml. Dilute 100 ml of this to 1000 ml with distilled water.

SPADNS solution

The color reagent
Dissolve 958 mg SPADNS in distilled water and dilute to 500 ml.

Zirconyl-acid reagent

The color bleaching reagent
Dissolve 133 mg zirconyl chloride octoahydrate, $ZrOCl_2.8H_2O$, in 25 ml distilled water. Add 350 ml concentrated HCl.
Dilute to 500 ml with distilled water.

Acid zirconyl–SPADNS

Mix equal volumes SPADNS and zirconyl acid reagent.

Hydrochloric acid

HCl, concentrated

Sodium arsenite

$NaAsO_2$
Dechlorinating agent
Dissolve 5 g $NaAsO_2$ and dilute to 1 l.

Procedure

1. Prepare 50 ml each of working standards (NaF) at concentrations from 1 to 2 mg/l.
2. Add 5 ml acid zirconyl-SPADNS reagent to each standard and to samples.
3. Set spectrophotometer to 0 absorbance or 100% transmittance with distilled water.
4. Read standards and samples on instrument at 570 nm and record.
5. Prepare calibration curve:

Using absorbance mode:

Set mg/l across the bottom of graph. Set absorbance along the left side with 100% at bottom, going up toward 0 absorbance. Plot standards; draw slope line.To calculate slope number on the absorbance scale, **measure the distance from the origin (100% Abs.) to the plotted point on the Y axis**.

To calculate sample concentration: Conc. = Abs × Slope Number. Again, measure the distance from the origin to the absorbance point of the sample.

Using transmittance mode:

Set mg/l across the bottom of graph. Set transmittance along the left side with 100% at bottom, going up toward 0 transmittance. Plot standards and draw slope line.

If the slope of the plotted standards is too high up on the graph, extend the line to the left until it crosses the Y axis, no matter where that point is. Don't try to force it to the corner at 100% transmittance. A new origin point has been created; use it. Measure the distance from that point to the plotted points on the Y axis. Use that distance in the slope calculations. Do the same with the sample point.

Analysis: Ion-Specific Electrode Method for Fluoride

With this method the sample interferences are eliminated by using a buffer, TISAB (Total Ionic Strength Adjustment Buffer), which contains CDTA, a chelating agent. The buffer also maintains the total ionic strength of the fluorides by breaking up covalent complexes, which fluorides tend to create at lower pHs. Two electrodes are used: a reference electrode and a fluoride electrode that senses ionic activity of the fluoride species. A combination electrode is also available.

Quality Control

- There is no need to distill the sample, as aluminum does not interfere with this method.
- Fluoride etches glass. Store concentrated solutions in plastic bottles.
- Electrode shelf life is 1 year. Replace annually.
- Check probe accuracy according to manufacturer's directions for millivolt change per decade of concentration.

Apparatus

- pH meter that reads out in millivolts
- Fluoride electrode (combination or fluoride electrode, plus reference electrode)

Reagents

Sodium fluoride

NaF, 10 mg/l
The standard
Dissolve 221 mg anhydrous sodium fluoride in distilled water and dilute to 1000 ml. Dilute 100 ml of this to 1000 ml with distilled water.

TISAB

Buffer, purchased
This solution is available as TISAB II. Use half TISAB and half sample, or TISAB III. Use 10 ml TISAB in 90-ml sample.

Procedure

1. Prepare fluoride standards at 0.1 mg/l, 1 mg/l, and 10 mg/l.
2. Measure a convenient amount of standards and samples into small beakers. Add TISAB (see Reagents).
3. Immerse electrodes in each standard and sample; wait for a stable reading. Withdraw electrode and blot dry between readings. Record your finding.
4. Plot on 4-cycle semilog graph paper.

Federal Limits

4 mg/l (primary MCL); 2 mg/l (secondary MCL).

References

American Public Health Association, American Water Works Association, and Water Environment Federation, *Standard Methods for the Examination of Water and Wastewater*, 20th ed., American Public Health Association, Washington, D.C., 1998, p. 4-79.

American Water Works Association, *Basic Science Concepts and Applications*, 2nd ed., American Water Works Association, Denver, CO, 1995, p. 511.

Chapter 21

Sulfate

Contents

Sulfate is one of the major anions occurring in natural waters. It is of importance in public water supplies because of its cathartic effect upon some humans when present in excessive amounts. Well waters high in hydrogen sulfide have a rotton egg or "sulfur" odor, especially noticeable when you take a hot shower. Sulfate in the water exists in equilibrium with hydrogen sulfide, a soluble odorous gas.

$$SO_4^{2+} \leftrightarrow H_2S$$

The direction the equilibrium shifts in depends somewhat on available oxygen in the water, and on pH (lower pH, more H_2S). The presence of sulfate-reducing bacteria is an important factor. These anaerobic bacteria derive their oxygen needed for metabolism by stripping it from sulfates. Sulfate may occur in increased concentration due to industrial discharge contamination from tanneries, paper mills, and industries that use sulfates and sulfuric acid in their processes. Mine drainage wastes are also high in sulfate. Acid rain can increase the sulfate concentration of the water. Excessively high concentrations of sulfate may decrease the pH of the water.

Treatment Plant Significance

Sulfate, only partially soluble when combined with calcium, produces hard scales in industrial boilers, heat exchangers, and in distribution piping. Conventional water and wastewater treatment processes cannot remove sulfate unless special treatment has been installed to do so. Hydrogen sulfide can be removed by scrubbing, or it can be converted to sulfate by aeration or chlorination. Decreased pH may be noted in waters with high sulfate concentrations and low alkalinity. pH adjustment should be made to prevent unwanted metals from dissolving.

> **NOTE: Hydrogen sulfide is a toxic gas. It may occur in great enough concentrations in a confined space to be a serious hazard.**

Analysis: Turbidimetric Method for Sulfate

Sulfate ion is converted to a predictable amount of barium sulfate suspension under controlled conditions. This turbidity is then read on a spectrophotometer and compared to sulfate standards.

$$BaCl_2 + SO_4^{2+} \rightarrow \underset{\text{suspension}}{BaSO_4} + 2Cl^-$$

Quality Control

- Water turbidity will interfere with this test. Filter turbid waters.
- Set standard range below 10 mg/l to produce a linear slope. Dilute the sample to fall within this range, if necessary.
- Pay special attention to timing and temperatures.
- Barium chloride is used in excess. Make sure enough has been added. If too much is added, it will not dissolve. When pouring aliquot into sample cell for spectrophotometer reading, omit these particles.

Apparatus

- Spectrophotometer, set at 420 nm
- Heater/mixer
- Stirbar

Reagents

Buffer solution A

Controls pH

For use when expected sulfate concentration is greater than 10 mg/l.
Dissolve 30 g $MgCl_2 \cdot 6H_2O$, 5 g $CH_3COONa \cdot 3H_2O$, 1 g KNO_3 and 20 ml CH_3COOH in 500 ml distilled water. Dilute to 1 l.

Buffer solution B

Controls pH
For use when expected sulfate concentration is less than 10 mg/l.
Dissolve 30 g $MgCl_2 \cdot 6H_2O$, 5 g $CH_3COONa \cdot 3H_2O$, 1 g KNO_3, 0.111 g Na_2SO_4, and 20 ml CH_3COOH in 500 ml distilled water. Dilute to 1 l.

Barium chloride

$BaCl_2$
Produces the turbidity
Crystals, 20 to 30 mesh

Standard sulfate solution

The standard
100 mg/l
Use either sulfuric acid or sodium sulfate.

Sulfuric acid

Dilute 10.4 ml standard 0.02 *N* H_2SO_4 with distilled water. See total alkalinity test for preparation.

Sodium sulfate

Na_2SO_4
Dissolve 0.1479 g anhydrous Na_2SO_4 in distilled water and dilute to 1 l.

Procedure

1. Prepare working standards in the range 1 to 10 mg/l. Dilute the sample to come within this range.
2. Measure 100-ml sample or a dilution made up to 100 ml into a flask or beaker.
3. Add 20 ml buffer. While stirring, add a spoonful $BaCl_2$ crystals. Let stir for 60 sec.
4. Read standards and samples on spectrophotometer.
5. Prepare standard curve. Calculate slope and sample concentration.

NOTE: Add chemical; mix, time, and read each sample separately so that timing will be the same for each.

Federal Limits

250 mg/l (secondary MCL).

Reference

American Public Health Association, American Water Works Association, and Water Environment Federation, *Standard Methods for the Examination of Water and Wastewater*, 20th ed., American Public Health Association, Washington, D.C., 1998, p. 4-176.

Chapter 22

Dissolved Oxygen

Contents

Dissolved oxygen (DO) is one of the most important and useful water measurements. Although the oxygen concentration in air is about 21%, in water it is only slightly soluble. Oxygen saturation ranges from 7 mg/l in hot water to 15 mg/l in cold water and is 9.2 mg/l at 20°C and atmospheric pressure at sea level. The type of life in a natural water body will depend upon the amount of DO present. Most microorganisms use free or DO for respiration. On the other hand, plant photosynthesis adds dissolved oxygen to the water as a by-product.

Treatment Plant Significance

Dissolved oxygen is most important in potable water systems for its effect on other chemicals in the water. It oxidizes both organics and inorganics, altering their chemical and physical states and their capacity as a nuisance to the customer. Corrosion of distribution system piping and appurtenances is enhanced by oxygen presence, and the measurement of DO is the basis for the BOD test in wastewaters.

Analysis: Winkler Procedure for Dissolved Oxygen (Azide Modification)

Oxygen cannot be determined directly by chemical methods. The Winkler procedure is an indirect method which is dependent upon the fact that dissolved oxygen oxidizes manganese ions (Mn^{2+}) to a tetravalent state (Mn^{4+}) under alkaline conditions. This more reactive state of manganese is capable of oxidizing iodine ions to free iodine under acid conditions. The amount of free iodine released is equivalent to the amount of oxygen originally in the sample. The iodine is then measured by reaction with standardized sodium thiosulfate prepared in a concentration such that 1 ml sodium thiosulfate = 1 mg/l DO.

Chemical Reactions — Winkler Procedure

$MnSO_4 + 2KOH \rightarrow Mn(OH)_2 + K2SO_4$	Azide reagent added. Mn^{2+} reacts with KOH and forms hydroxide floc.
$2Mn(OH)_2 + O_2 \rightarrow 2MnO(OH)_2$	Reaction proceeds. Hydroxide reacts with O_2; Mn takes tetravalent state.
$MnO(OH)_2 + 2H_2SO_4 \rightarrow Mn(SO_4)_2 + 3H_2O$	Sulfuric acid added. Floc dissolves.
$Mn(SO_4)_2 + 2KI \rightarrow MnSO_4 + K_2SO_4 + I_2$	Mn reacts with the KI from azide reagent. Tetravalent charge is released and free iodine (yellow) is produced.
$I_2 + 2Na_2S_2O_3 \rightarrow Na_2S_4O_6 + 2NaI$	Titrating iodine discharges yellow color. Starch is added for endpoint recognition.

Quality Control

- Measure DO immediately after taking sample (on site if possible).
- Do not shake sample.
- Do not change temperature.
- Do not dilute sample.
- Do not let air in while sampling or measuring.
- The Winkler procedure is best suited to clean waters. The Azide modification eliminates interference from nitrites, but dissolved organics, suspended solids, and iron will interfere with this test.
- Ideally, DO testing should be done on site, at the sampling location. Care must be taken in collecting samples. Turbulence will put extra oxygen into an undersaturated sample. Sample bottles should be filled to the top and sealed so that a change in temperature will not affect the amount of oxygen in the sample.

- It is difficult to obtain accurate oxygen concentration of well waters. Pumping equipment may entrain oxygen and add to its concentration in the sample.
- Starch supports bacterial growth. The shelf life is 1 month at best, unless a preservative is added.

Apparatus

- BOD bottles

Reagents

Manganous sulfate solution

$MnSO_4$
Oxidized by DO to tetravalent state, Mn^{4+}
Dissolve 480 g $MnSO_4{\cdot}4H_2O$ in distilled water. Filter and dilute to 1 l.

Alkali-iodide-azide reagent

Alkali is reduced to provide tetravalent anion for the manganese. Iodide provides the iodine, which is directly measured, and azide eliminates nitrite interference. Dissolve 500 g NaOH and 135 g NaI in distilled water and dilute to 1 l. Dissolve 10 g NaN_3 in 40 ml distilled water and add to above solution. Caution: dissolution of NaOH is exothermic and should be done in an ice bath.

Starch

Titration endpoint indicator
Add 5 g starch to 1 l boiling distilled water. Let stand overnight to settle, decant supernate, and use.

Standard sodium thiosulfate

$Na_2S_2O_3{\cdot}5H_2O$
The titrant
Dissolve 6.205 g $Na_2S_2O_3{\cdot}5H_2O$ in distilled water. Add 0.4 g solid NaOH and dilute to 1 l. *Must be standardized.*

Potassium bi-iodate

$KH(IO_3)_2$, 0.0021 *M*
Primary standard, to standardize the sodium thiosulfate
Dissolve 812.4 mg $KH(IO_3)_2$ in distilled water and dilute to 1 l.

Sulfuric acid, concentrated

H_2SO_4
Dissolves the tetravalent manganese floc

Procedure

1. Standardize the sodium thiosulfate:

   Dissolve 2 g KI in an Erlenmeyer flask with 100 ml distilled water.

   Add a few drops concentrated H_2SO_4 and exactly 20 ml potassium bi-iodate solution.

   Dilute to 200 ml and titrate iodine (yellow) with sodium thiosulfate titrant.

   When near end of titration (pale straw color) add starch (turns blue) and continue titrating endpoint.

   Adjust sodium thiosulfate to 0.025 *M*.
2. Fill 300 ml BOD bottle to the top with sample.
3. Add (submerged) 1 ml manganous sulfate solution and 1 ml alkali-iodide-azide reagent. Stopper, excluding bubbles, and mix by inverting a few times.
4. Let precipitate settle to half way down bottle. Add 1 ml concentrated H_2SO_4 and mix. Wait until it dissolves.
5. Remove 200 ml and titrate with sodium thiosulfate. Near end of titration (straw color) add starch and continue titration to the endpoint. Starch, an absorption indicator added at the end of the titration, forms a blue complex with the remaining iodine and provides better visibility of the endpoint.
6. Calculation: 1 ml 0.025 *M* $Na_2S_2O_3$ = 1 mg/l DO.

Analysis: Membrane Electrode Method for Dissolved Oxygen (DO meter)

The electrode method for measuring DO has widespread use because it eliminates interferences and allows the chemist to take measurements on site. The instrument is a milliammeter with an electrode probe on a cord. An oxygen permeable membrane is stretched across the end of the probe, and holds in the sensing electrode, which is submerged in an anoxic electrolyte solution. The probe fits snugly into the neck of a BOD bottle filled with sample. A small stirrer is attached to the end of the probe, which keeps new water flowing past the membrane as it reads. Differential pressure makes oxygen molecules pass through the membrane where they are reduced (take on electrons) at a cathode, and then flow to an anode, oxidizing it. The result is a flow of electrons from cathode to anode proportional to the oxygen passing through the membrane. The electrical signal is then converted to concentration units which is read on the meter. Temperature is important, and is monitored by a thermistor built into the probe.

Quality Control

- A DO meter must be calibrated. For daily use it can be calibrated to the oxygen saturation point (wet probe in air), but it should also periodically be calibrated to the Winkler procedure results for the same sample.

- Probe membranes, or caps should be changed frequently, according to manufacturer's directions and whenever readings are unusual. Suspended solids collecting on the membrane will interfere with the electrode operation.
- Read DO immediately after sampling.

Apparatus

- Dissolved oxygen meter and probe
- BOD bottle

Procedure

1. Calibrate DO meter, either by air calibration or by Winkler procedure.
2. Fill bottle to neck with sample.
3. Insert probe into neck of bottle. Do not allow any trapped air bubbles.
4. Turn on stirrer. Read dissolved oxygen.

Effect of Temperature on Oxygen Saturation (at 1 Atm Pressure)

°C	mg/l DO (sat)	°C	mg/l DO (sat)
0	14.6	13	10.6
1	14.2	14	10.4
2	13.8	15	10.2
3	13.5	16	10.0
4	13.1	17	9.7
5	12.8	18	9.5
6	12.5	19	9.4
7	12.2	20	9.2
8	11.9	21	9.0
9	11.6	22	8.8
10	11.3	23	8.7
11	11.1	24	8.5
12	10.8	25	8.4

Federal Limits

None.

Reference

American Public Health Association, American Water Works Association, and Water Environment Federation, *Standard Methods for the Examination of Water and Wastewater*, 20th ed., American Public Health Association, Washington, D.C., 1998, p. 4-129.

Chapter 23

Solids

Contents

Solids in water are defined as any matter that remains as residue upon evaporation and drying at 103°C. They are separated into two classes: suspended solids and dissolved solids.

Total Solids = Suspended Solids + Dissolved Solids
(nonfilterable residue) (filterable residue)

Each of these has volatile (organic) and fixed (inorganic) components that can be separated by burning in a muffle furnace at 550°C. The organic components are converted to carbon dioxide and water and the ash is left. Weight of the volatile solids can be calculated by subtracting the ash weight from the total dry weight of the solids.

Total Suspended Solids/Volatile Solids

The total suspended solids test is extremely valuable in the analysis of polluted waters, and is sometimes performed on clean waters as a check on quality.

Treatment Plant Significance

For naturally occurring waters whose quality is suspect or changing, the total suspended solids test can be useful when compared to turbidity trends, and to the results of the total and fecal coliform tests. This is not a routine test performed at potable water treatment facilities in most areas, but it has value in water quality monitoring. In treatment process, total suspended solids is equivalent to the turbidity and can be removed by the same mechanisms. However, there is no legal restriction on total suspended solids, as such.

Analysis: Total Suspended Solids/Volatile Solids

Known volumes of samples are filtered, dried, and weighed to determine the suspended solids concentration. Burning the residue in a muffle furnace yields the volatile (organic) content.

Quality Control

- Use graduated cylinder to measure sample volume. Transfer quantitatively to Gooch crucible. Wash out any particles that are stuck on the inner walls of the cylinder.
- Calibrate the analytical balance to standard weights to the milligram.
- Do not handle solids sample ceramics with fingers. Use tongs.
- Weigh only room temperature samples. Allow warm samples to cool in a desiccator.
- To troubleshoot weight discrepancies, run a blank using distilled water, along with the samples.

- As per *Standard Methods*, periodically dry, cool, and weigh repeatedly until weight is constant. In this way proper drying time can be established.
- Oven and furnace drying and burning temperatures are critical. A calibrated thermometer should be permanently set into the drying oven. It should be checked periodically with another calibrated thermometer.
- Don't allow the muffle furnace go over 550°C. Components in the filter paper may ignite. At higher temperatures, some inorganics in the sample may burn.
- Do not open the furnace door after a sample has been inserted for ignition. Oxygen entering when the door is opened may ignite the sample forcefully enough to blow some of it out of the crucible.

Apparatus

- Drying oven
- Muffle furnace
- Gooch crucibles
- Desiccator
- Glass fiber filters
- Source of vacuum
- Filtering flask

Reagents

None

Procedure

1. Insert filter with rough side up in Gooch crucible. Rinse filter with distilled water, applying vacuum to seat.
2. Dry filter and Gooch crucible at 103°C for 1 h. Burn in muffle furnace at 550°C for 15 min. Cool in desiccator.
3. Weigh Gooch/filter to the nearest mg (tare weight).
4. Transfer sample quantitatively to crucible. Filter a known volume of sample. It is best to use the largest volume possible that will not blind the filter.
5. Dry in 103°C oven for 1 h. Cool in desiccator. Weigh again (dry weight).
6. Burn in muffle furnace for 15 min at 550°C. Cool in desiccator and weigh again (ashed weight).
7. Calculate:

$$\textbf{TSS (mg/l)} = \frac{\textbf{Dry Weight (mg) – Tare Weight (mg)} \times \textbf{1000}}{\textbf{ml sample}}$$

$$\text{Fixed Solids (mg/l)} = \frac{\text{Ashed Weight (mg)} - \text{Tare Weight (mg)} \times 1000}{\text{ml sample}}$$

$$\text{Volatile Solids (mg/l)} = \text{TSS (mg/l)} - \text{Fixed Solids (mg/l)}$$

$$\%\text{Volatile Solids} = \frac{\text{Volatile Solids (mg/l)} \times 100}{\text{TSS (mg/l)}}$$

> NOTE: **Use care with heating equipment. A muffle furnace set at burning temperature is over 1000°F!**

Federal Limits

None.

Reference

American Public Health Association, American Water Works Association, and Water Environment Federation, *Standard Methods for the Examination of Water and Wastewater*, 20th ed., American Public Health Association, Washington, D.C., 1998, p. 2-57.

Total Solids/Volatile Solids

Total solids tests can be performed on any liquid sample and will include both suspended and dissolved solids. The result may be registered in mg/l or as a percentage. The test is occasionally used in general monitoring of water quality, keeping in mind that in some clean waters, much of the total solids content may be due to large concentrations of dissolved components such as hardness and alkalinity ions.

Treatment Plant Significance

There is no legal regulation regarding total solids in potable water. Removal in conventional treatment is specific to the treatment process. Coagulation, settling, and filtration removes suspended solids, while softening removes dissolved solids.

Analysis: Total Solids/Volatile Solids

In the total solids test, known volume is generally not taken. Instead, an evaporating dish or other burnable container is filled and wet weight, dry weight, and ash weight are recorded. Percent of total solids can then be determined.

Quality Control

- For water, rather than sludge samples, dry at 98°C until the visible water is gone, to prevent spattering. Then turn drying oven up to 103°C.
- Mix the sample well. Pour aliquot before solids get a chance to settle. Transfer quantitatively to evaporating dish. Wash out any particles that are stuck on the inner walls of the cylinder. If large, uncharacteristic pieces are in the sample, remove them.
- Calibrate the analytical balance to standard weights to the milligram.
- Do not handle solids sample ceramics with fingers. Use tongs.
- Weigh only room temperature samples. Allow warm samples to cool in a desiccator.
- To troubleshoot weight discrepancies, run a blank using distilled water, along with the samples.
- As per *Standard Methods*, dry, cool, and weigh repeatedly until weight is constant. In this way, proper drying time can be established.
- Oven and furnace drying and burning temperatures are critical. A calibrated thermometer should be permanently set into the drying oven. Periodically it should be checked with another calibrated thermometer.
- For low total solids waters, the sample should be as large as possible to assure maximum accuracy. If a volatile solids test is not to be performed, a beaker can be used to hold the sample.

Apparatus

- Drying oven
- Muffle furnace
- Desiccator
- Evaporating dish

Procedure

1. Burn clean evaporating dish in muffle furnace for 15 min at 550°C. Cool. Weigh dish to the nearest milligram (tare weight).
2. Fill dish about 3/4 full with sample. Weigh again (wet weight).
3. Dry sample in 103°C oven overnight, or until apparently dry. Cool in desiccator. Weigh again (dry weight).
4. Burn in muffle furnace at 550°C. Cool in desiccator. Weigh again (ashed weight).
5. Calculate:

$$\textbf{Total Solids (\%)} = \frac{\textbf{Dry Weight (g)} - \textbf{Tare Weight (g)} \times \textbf{100}}{\textbf{Wet Weight (g)} - \textbf{Tare Weight (g)}}$$

$$\textbf{Total Solids (mg/l)} = \textbf{Total Solids (\%)} \times \textbf{10,000}$$

$$\textbf{Volatile Solids (\%)} = \frac{\textbf{Ashed Weight (g) – Tare Weight (g)} \times \textbf{100}}{\textbf{Wet Weight (g) – Tare Weight (g)}}$$

$$\textbf{Volatile Solids (mg/l)} = \textbf{Volatile Solids (\%)} \times \textbf{10,000}$$

Federal Limits

None.

Reference

American Public Health Association, American Water Works Association, and Water Environment Federation, *Standard Methods for the Examination of Water and Wastewater*, 20th ed., American Public Health Association, Washington, D.C., 1998, p. 2-55.

Total Dissolved Solids

Total dissolved solids (TDS) are those that remain after the sample has been filtered and the suspended solids removed. Most of the solids concentration in clean natural waters is inorganic dissolved solids, and major components are salts of calcium, magnesium, and sodium.

Treatment Plant Significance

There is a secondary MCL set for total dissolved solids (TDS) at 500 mg/l. The results of this test do not stand alone, but relate to hardness, alkalinity, scale-forming tendency, corrosivity, and conductivity.

Quality Control

- Samples high in hardness and alkalinity may require prolonged drying time. Some may form a water-trapping crust; if this problem is suspected, limit sample size to 200-mg solids.
- Weigh sample, dry. Weigh repeatedly until constant weight is obtained.

Apparatus

- Drying oven
- Gooch crucibles

- Glass fiber filters
- Evaporating dish
- Desiccator
- Filtering flask
- Source of vacuum

Procedure

1. Filter sample and wash three times with small volume of distilled water.
2. Transfer filtrate to a weighed evaporating dish. Dry in oven at 180°C.
3. Cool in desiccator; weigh.
4. Calculate:

$$\textbf{Total Dissolved Solids/l} = \frac{\textbf{(Weight Dried Solids + Dish)} - \textbf{Weight Dish} \times \textbf{1000}}{\textbf{Sample Volume}}$$

Federal Limits

500 mg/l (secondary MCL).

Reference

American Public Health Association, American Water Works Association, and Water Environment Federation, *Standard Methods for the Examination of Water and Wastewater*, 20th ed., American Public Health Association, Washington, D.C., 1998, p. 2-56.

Chapter 24

Nitrogen

Contents

The compounds of nitrogen are of great interest to the water chemist because of the importance of nitrogen in life's processes. The chemistry of nitrogen is complex. This element can take seven oxidation states, five of which are of direct interest: nitrogen gas (N_2), organic nitrogen (C-C-C-NH_2), ammonia (NH_3), nitrite (NO_2^-), and nitrate (NO_3^-). The atmosphere is the earth's source of nitrogen, where it exists as the elemental gas and is constantly removed and recycled for life's needs. Electrical discharge and nitrogen-fixing bacteria make atmospheric nitrogen available to plants for growth. Animals in turn consume and break down plant protein for their own needs. The decomposition of plant and animal matter produces ammonia, which is oxidized by nitrifying bacteria (nitrosomonas and nitrobacter) to nitrate for plant use.

Synthetic fertilizers contain ammonium and nitrate compounds. The soil does not hold these materials well, and nitrates applied in excess of plant needs percolate to groundwater. Under anaerobic conditions, nitrates are reduced to nitrogen gas, which escapes back to the atmosphere.

It was common years ago for chemists to use tests for nitrogen to determine the sanitary quality of waters and the strength of wastewaters. Freshly polluted waters can have high levels of ammonia, which is gradually oxidized by bacteria to nitrite, then to nitrate. Where sludges are applied to land, both ammonia and nitrate are monitored to determine crop uptake and prevent leaching to groundwater.

In environmental testing, nitrogen is split into three categories: organic nitrogen (the sum of the concentrations of these two is called total Kjeldahl nitrogen), ammonia nitrogen, and inorganic nitrogen (nitrites and nitrates).

Nitrogen concentrations are designated in terms of the weight of nitrogen atoms in the sample, and the concentration is stated this way regardless of which form of nitrogen is being considered.

$$10 \text{ mg/l } NO_3\text{, as N} \quad \text{or} \quad 10 \text{ mg/l } NO_3\text{-N}$$

This means that the ion being measured is nitrate, but every liter of this solution has 10 mg of nitrogen in it. The weight of the oxygen is not counted.

Organic Nitrogen

Organic nitrogen is the sum of all the nitrogen present in organic compounds: amines, amides, amino acids, nitro derivatives, and others. The test procedure that digests organic nitrogen is the initial step of the total Kjeldahl nitrogen test.

Treatment Plant Significance

Organic nitrogen usually occurs in only very small concentrations in most waters. Its domestic source in wastewaters is urea, a major compound of urine, most of which is converted to free ammonia in wastewater collection systems. Industrial wastewaters may contribute significant amounts of organic nitrogen (protein wastes from food processing and synthetics from the fabric industry) but in clean surface waters the source is runoff. The test for total Kjeldahl nitrogen or organic nitrogen may be required for industrial processing.

Analysis: Macro-Kjeldahl Digestion Method for Organic Nitrogen

Samples are digested with concentrated sulfuric acid and a catalyst to break up the organic compounds, releasing the nitrogen as ammonia. The procedure is done with exhaust apparatus to vent toxic sulfite fumes which are produced. The sample is then transferred to a distillation flask and the ammonia is distilled off. A titration or colorimetric procedure is performed to measure the resulting total free ammonia concentration (total Kjeldahl nitrogen).

The organic nitrogen concentration is the difference between the total Kjeldahl nitrogen result, and the result of a test for free ammonia only, in the original sample.

Quality Control

- If colorimetric measurement is to be done, run standards and blank through digestion along with sample.
- An excess of 10 mg/l nitrate, a large quantity of dissolved salt, inorganic solids, or organics may interfere with the digestion. To avoid this, be sure that there is at least 1 ml sulfuric acid added per g of salt in the sample (or 1 ml sulfuric acid per every 3 g COD in the sample).

Apparatus

- Kjeldahl flasks, 800 ml
- Heating unit capable of 370°C with fume ejection apparatus

Reagents

Mercuric sulfate solution

$HgSO_4$
Digestion catalyst
Dissolve 8 g red mercuric oxide in 10 ml 6 *N* H_2SO_4.

Digestion reagent

Dissolve 134 g K_2SO_4 in 650 ml water and 200 ml concentrated H_2SO_4. Stirring, add 25 ml $HgSO_4$ solution. Dilute to 1 l.

Sodium hydroxide–sodium thiosulfate reagent

Neutralizes digestion reagent and raises pH for distillation
Dissolve 500 g NaOH and 25 g $Na_2S_2O_3 \cdot 5H_2O$ in water. Dilute to 1 l.

Borate buffer solution

Neutralizes strong acids in sample
Add 88 ml 0.1 *N* NaOH solution to 500 ml 0.025 *M* sodium tetraborate ($Na_2B_4O_7$) solution. Dilute to 1 l.

Sodium tetraborate solution

$Na_2B_4O_7 \cdot 10H_2O$, 0.025 *M*
For borate buffer
Dissolve 9.5 g $Na_2B_4O_7 \cdot 10H_2O$ in water. Dilute to 1 l.

Sodium hydroxide

NaOH, 6 *N*
Converts ammonia in sample to gaseous state
Dissolve 240 g NaOH in water. Dilute to 1 l.

Procedure

1. Select sample volume from the following:

Organic Nitrogen, mg/l	Sample Volume, ml
0–1	500
1–10	250
10–20	100
20–50	50
50–100	25

2. Bring sample up to 500 ml with distilled water. Add 25 ml borate buffer and then 6 *N* NaOH until pH is 9.5. Add boiling chips and boil off 300 ml to drive off free ammonia and cool.
3. Carefully add 50 ml digestion reagent. Add boiling chips; mix. Attach to ejection equipment and heat until volume is about 30 ml and copious white fumes are observed. Digest for 30 min more and cool.
4. Dilute to 300 ml with water and mix.
5. Carefully add 50 ml hydroxide-thiosulfate reagent to form an alkaline layer at flask bottom. Connect to distillation apparatus. Shake and distill off free ammonia.

NOTE: Use care when adding digestion reagents. These are concentrated solutions. Violent reactions can occur.

Federal Limits

None.

Reference

American Public Health Association, American Water Works Association, and Water Environment Federation, *Standard Methods for the Examination of Water and Wastewater*, 20th ed., American Public Health Association, Washington, D.C., 1998, 4-123.

Ammonia Nitrogen

Ammonia is a product of the microbiological decay of animal and plant protein. When dissolved in water at a pH of 7 or less, almost all of it occurs as the ammonium ion (NH_4^+). As the pH rises, it converts to the unionized gaseous form (NH_3). In any water, the two forms of ammonia are in equilibrium:

$$NH_4^+ \leftrightarrow NH_3$$

Ammonium ion is a potential nutrient to microorganism life; upon oxidation it stimulates the growth of bacteria and algae. Unionized ammonia (the gas) is toxic to many organisms; even tiny quantities can be detrimental to life. A clean, natural waterway will have many aerobic nitrifying bacteria that will act on ammonia, converting it to nitrate.

Treatment Plant Significance

When chlorine is added to any water that has ammonia present — whether it is a trace amount of ammonia in a clean potable water source, or a great deal of ammonia in a wastewater discharge — the chlorine will react with the ammonia to form chloramines (combined chlorine residual), which is an effective and long-lasting disinfectant. To provide free chlorine disinfection the utility must add enough chlorine to overcome this reaction, break up the chloramines, and convert all added chlorine to free chlorine (breakpoint chlorination). If difficulty is encountered in achieving this breakpoint, an analysis for ammonia may be considered.

Analysis: Preliminary Distillation of Free Ammonia Nitrogen

When ammonia concentration is more than 5 mg/l and colorimetric or titrimetric measurement is to be used, the sample must be distilled first. The free ammonia is collected in a solution of boric acid (as ammonium borate). This eliminates any interfering ions, color or turbidity.

Quality Control

- When testing for nitrogen components, always use distilled, deionized water to prepare reagents and standards. These ions are often present in the sample in very small amounts, and traces of nitrogen in the distilled water may lead to inaccurate results.
- Run standards and blank through distillation along with sample.

Apparatus

- Distillation apparatus
- pH meter
- Deionizer

Reagents

Ammonia-free distilled water

Pass distilled water through deionizer column appropriate for removing ammonia and store in a tightly stoppered glass container. Use for preparing reagents, standards, and for making dilutions.

Borate buffer solution

$Na_2B_4O_7$

Add 88 ml 0.1 *N* NaOH solution to 500 ml 0.025 *M* sodium tetraborate ($Na_2B_4O_7$) and dilute to 1 l.

Sodium tetraborate solution

$Na_2B_4O_7 \cdot {}_{10}H_2O$, 0.025 *M*

For borate buffer

Dissolve 9.5 g $Na_2B_4O_7 \cdot {}_{10}H_2O$ in water. Dilute to 1 l.

Sodium sulfite

$NaSO_3$

Dechlorinating agent

Dissolve 0.9 g $NaSO_3$ in water and dilute to 1 l. Prepare fresh daily (1 ml neutralizes 1 mg/l chlorine).

Sodium hydroxide

NaOH, 6 *N*

Raises pH to change ammonium ion to gaseous ammonia

Dissolve 240 g NaOH in water and dilute to 1 l.

Boric acid

H_3BO_3

Receiving solution, to collect ammonia distillate as ammonium borate

Dissolve 20 g boric acid (H_3BO_3) in water and dilute to 1 l.

Indicating boric acid solution

For titrimetric method of measurement, to indicate endpoint

Prepare boric acid solution. Add 10 ml mixed indicator solution. Prepare monthly.

Mixed indicator solution

For titration endpoint

Dissolve 200 mg methyl red in 100 ml 95% ethyl alcohol. Dissolve 100 mg methylene blue in 50 ml 95% ethyl alcohol and combine solutions. Prepare monthly.

Procedure

1. Prepare distillation equipment. Add 500 ml water and 20 ml borate buffer to each flask. Adjust pH to 9.5 with 6 *N* NaOH solution. Add boiling chips and boil for 30 min to distill residual ammonia from flasks. Discard solution.
2. Dechlorinate sample with sodium sulfite, if necessary.
3. Put 500 ml sample in distillation flask. Standards and blank should also be run through distillation. Add 25 ml borate buffer and adjust pH to 9.5 with 6 *N* NaOH.
4. Put 50 ml boric acid solution (indicating boric acid solution if doing acid titration) in a 500-ml Erlenmeyer flask and submerge end of distillation tube in this solution.
5. Distill standards, blank, and sample until 300 ml distillate have been collected in receiving solution. Remove tube from solution and turn still off.
6. Bring standards, blank, and sample back to 500 ml with distilled water.
7. Measure ammonia by Nesslerization, acid titration, or ion electrode method.

NOTE: **When distillation is complete, withdraw submerged tube from Erlenmeyer flask immediately. Backsuction will occur and will draw the distillate right back into the distillation flask.**

Analysis: Nesslerization Determination Method for Free Ammonia Nitrogen

The direct Nesslerization procedure is meant for clean waters only and works best on samples with an ammonia concentration of less than 0.5 mg/l. If interfering ions are present (ketones, aldehydes, alcohols, amines, calcium, or magnesium), distill first. The sample is pretreated with zinc sulfate to settle out interfering ions. Then Nessler reagent, a strongly alkaline solution of potassium mercuric iodide, is added to the sample and to a set of ammonium standards. A yellow-brown color is developed and comparison is made to standards by spectrophotometer.

Quality Control

- Warm up pH meter 30 min before using. Calibrate with two buffers.
- Samples to be tested for ammonia will change upon standing. Refrigerate samples to inhibit bacterial growth, add sulfuric acid to pH 2, and perform analysis within 24 h.
- Accuracy of standards is of prime importance. Correct calculation of sample concentration depends upon this.

- Use small concentration standards (1, 2, 3, or 4 ppm), or the standard curve will lose linearity. Dilute the sample and adjust calculation for the dilution later.
- Check results of one method against results of another.
- To check the distillation, run the standards and blank through this part of the procedure along with the samples.
- Keep Nessler reagent in the dark. If it develops a precipitate, prepare a new reagent.

Apparatus

- Spectrophotometer
- pH meter

Reagents

Nessler reagent

Produces a yellow-brown color when reacted with ammonia nitrogen

Dissolve 100 g HgI_2 and 70 g KI in a small quantity of distilled water. Add this mixture, while stirring, to a cool solution of 160 g NaOH dissolved in 500 ml water. Dilute to 1 l and store in a dark bottle. Solution is stable for 1 year.

Stock ammonium solution

NH_4Cl

The standard

Dissolve 3.819 g anhydrous NH_4Cl (dried, 100°C) in water and dilute to 1 l; (1000 mg/l). Make dilutions as needed for working standards from this stock. Prepare fresh standard every 2 weeks.

Zinc sulfate solution

$ZnSO_4{\cdot}7H_2O$

Pretreatment for undistilled samples, settles out interfering ions

Dissolve 100 g $ZnSO_4{\cdot}7H_2O$ in water and dilute to 1 l.

Sodium hydroxide

NaOH, 6 *N*

Raises pH to convert ammonium ion to gaseous ammonia

Dissolve 240 g NaOH in water and dilute to 1 l.

EDTA reagent

Holds any remaining Ca, Mg ions in solution to eliminate turbidity interference with Nessler reagent

Dissolve 10 g NaOH in 60 ml distilled water. Dissolve 50 g EDTA (disodium salt) into this and warm slightly to dissolve.

Procedure

1. If sample is undistilled, pretreat:

   Dechlorinate sample; if sample is expected to have >5 mg/l, dilute

   Add 1 ml $ZnSO_4$ solution to 100-ml sample and mix.

   Add 6 *N* NaOH to obtain a pH of 10.5. Mix and let stand for a few minutes.

   After floc settles, filter (discard first 25 ml filtrate) or centrifuge, saving 50 ml filtrate.

   Add 1 drop EDTA reagent and mix.
2. Add 2 ml Nessler reagent to sample, standards, and blank. Mix and let stand for 10 to 30 min for color development. Be sure that time, temperature, etc. are the same for all standards, blank, and sample.
3. Read on spectrophotometer at 400 to 450 nm. Prepare calibration curve and calculate sample concentration.

Analysis: Acid Titration for Free Ammonia Nitrogen

This is an ammonia determination method which is used for samples which already have been distilled. The ammonia distillate is collected in indicating boric acid and titrated to the original pH of the acid. This neutralizes the borate which is tied up with the ammonia in solution, thus measuring the amount of ammonia there.

Quality Control

- A blank and standards should be carried through the distillation and titrated.

Apparatus

- pH meter

Reagents

Sodium carbonate

Na_2CO_3, 0.05 *N*
To standardize the sulfuric acid
Dry 5 g Na_2CO_3 at 250°C for 4 h. Cool in desiccator. Dissolve 2.5 g in water and dilute to 1 l.

Sulfuric acid

H_2SO_4, 0.02 *N*
Titrant
Mix 2.8 ml H_2SO_4 with water. Dilute to 1 l. Mix well and titrate into 40 ml 0.05 *N* sodium carbonate solution using pH meter to pH 5. Boil gently 5 min under watch glass to remove CO_2. Cool to room temperature and finish titrating to pH inflection point. Calculate normality of acid and dilute to 0.02 *N*.

Procedure

1. Titrate standard sulfuric acid into distillate until indicator turns pale lavender. Calculate ammonia concentration:

$$\text{mg/l NH}_3\text{-N} = \frac{(A - B) \times 280}{\text{ml sample}}$$

where: A = ml H_2SO_4 titrated for sample
B = ml H_2SO_4 titrated for blank

Analysis: Ammonia Selective Electrode Method for Free Ammonia Nitrogen

The ammonia electrode is composed of a combination pH electrode with a gas permeable membrane attached to its end to separate the sample from the electrodes internal filling solution. The sample is made basic so that all free ammonia changes into the gaseous state (pH 11). Dissolved ammonia gas will pass through the membrane theoretically until the partial pressure on both sides is the same. The internal filling solution, however, is acidic enough to change the gas to ammonium ion as soon as it passes through. The pH of the alkaline ammonia ion is then sensed by the pH electrode, and converted to a millivolt reading. Pressure is maintained on the outside, and the gas continues to pass through the membrane.

Ammonia standards must be used, most often, 1 mg/l, 10 mg/l, and 100 mg/l.

Quality Control

- Distillation is advised, but not necessary, if ammonia concentration is more than 5 mg/l and ion selective electrode is to be used for measurement. Amines, mercury, and silver interfere.
- To calibrate meter, the slope of the calibration curve (millivolts per decade of concentration) should be about 90. If it drops to near 50, change the filling solution and cap.

- Add NaOH and wait at least 3 to 4 min. It takes this long for the ammonia gas to build up.
- Specific-ion electrodes have a usable life of 1 year.
- Membrane caps should be replaced every month (with new filling solution). Erratic standard readings usually means that the membrane cap should be replaced.
- Store probe in diluted standard solution.

Apparatus

- pH meter
- Ammonia selective electrode
- Deionizer

Reagents

Ammonia-free distilled water

Pass distilled water through deionizer column appropriate for removing ammonia and store in tightly stoppered glass container. Use for preparing reagents, standards, and for making dilutions.

Stock ammonium solution

NH_4Cl, 1000 mg/l

The standard

Dissolve 3.819 g anhydrous NH_4Cl (dried, 100°C) in distilled water and dilute to 1 l. Make dilutions as needed for working standards from this stock.

Sodium hydroxide

NaOH, 10 *N*

Raise pH to convert ammonium ion to gaseous ammonia

Dissolve 400 g NaOH in 800 ml water. Cool and dilute to 1 l.

Procedure

1. From stock solution, prepare working standards of 100 mg/l, 10 mg/l, and 1 mg/l.
2. Immerse probe into each standard and sample on auto-stirrer. Add 1 ml NaOH, wait 5 min for conversion to take place, and read.
3. Construct standard curve on 4-cycle semilog graph paper; mg/l plotted against millivolts on arithmetic graph paper does not produce a linear curve. Use of the log paper is an attempt to straighten the line. Set up mg/l across bottom and millivolts up the side. Set the lowest standard at the X-Y intersection.

Federal Limits

None.

Reference

American Public Health Association, American Water Works Association, and Water Environment Federation, *Standard Methods for the Examination of Water and Wastewater*, 20th ed., American Public Health Association, Washington, D.C., 1998, p. 4-106.

Nitrite Nitrogen

The nitrite ion (NO_2^-) is most unstable and rarely found in more than trace concentrations in any water. This is the partially oxidized state that occurs briefly as ammonia is oxidized, or as nitrate is reduced, and normal concentrations are well under 1 mg/l.

Treatment Plant Significance

There is a mandatory limit on nitrite in potable waters (1 mg/l) because of its tendency to oxidize to nitrate, which has a detrimental health significance. Conventional potable water treatment plants are not able to remove either nitrite or nitrate, and the responsibility may be left to the upstream wastewater treatment plant as the source of excess nitrogen in the water.

Analysis: NED Diazotization Method for Nitrite Nitrogen

The NED diazotization procedure for nitrite is a sensitive and reliable colorimetric method that can detect nitrite in very small concentrations. Under acid conditions, nitrite reacts with NED reagent (N-(1-naphthyl)-ethylenediamine dihydrochloride) to form a wine-colored azo dye that is proportional to the nitrite concentration. Comparison can be done by colorimetry using nitrite standards.

Quality Control

- Care is needed in preparing standards. Mix well. Concentration to be measured is very small.
- Nitrite is unstable. Nitrite standards can be kept for only 1 day. Samples must be tested for nitrite immediately upon collection.

Apparatus

- Spectrophotometer
- Demineralizer

Reagents

Nitrite-free distilled water

Run distilled water through appropriate demineralizer column for removing nitrite. Use this water for all reagents, standards, and dilutions.

NED reagent

N-(1-naphthyl)-ethylenediamine dihydrochloride
Produces the color
Add 100 ml 85% phosphoric acid and 10 g sulfanilamide to 80 ml distilled water. When dissolved, add 1 g NED, mix, and dilute to 1 l with distilled water. Store in dark bottle. Reagent is stable for 1 month.

Standard potassium permanganate

$KMnO_4$, 0.05 *M*
Used to standardize the nitrite stock solution
Dissolve 8 g $KMnO_4$ in 1 l water. Store in brown bottle and age for at least 1 week. *Must be standardized.*

Nitrite stock solution

$NaNO_2$, 50 mg/l, 0.0179 *M*
The standard
Dissolve 1.232 g $NaNO_2$ in water and dilute to 1 l. Must be standardized. Prepare working standards from this solution. Make fresh daily.

Sodium oxalate

$Na_2C_2O_4$, 0.05 *N*
Primary standard; to standardize the permanganate
Dissolve 3.35 g $Na_2C_2O_4$ in water and dilute to 1 l.

Sulfuric acid

H_2SO_4, 1+1
Add 20 ml concentrated sulfuric acid to 20 ml distilled water and mix.

Procedure

1. Standardize potassium permanganate:

   Dissolve 200 mg sodium oxalate in 100 ml water.

   Add 10 ml 1+1 H_2SO_4 and heat to 95°C. Keep heat above 85°C through entire procedure.

   Titrate permanganate solution into this mixture until faint pink color appears and persists for 1 min.

   Calculate: $$\text{Molarity } KMnO_4 = \frac{g\ Na_2C_2O_4}{\text{ml titrated} \times 0.335}$$

   Adjust to 0.05 *M*.
2. Standardize nitrite stock solution:

   Add, in this order, 50 ml standard 0.05 *M* $KMnO_4$, 5 ml concentrated H_2SO_4, and 50 ml stock nitrite solution.

   Warm to 80°C on hot plate.

   Titrate $Na_2C_2O_4$ into this solution until permanganate color is just barely discharged.

   Calculate molarity of stock solution.
3. Prepare working standards 0.2 to 1.0 mg/l, or expected range. Prepare 50 ml of each.
4. Filter sample to remove turbidity.
5. Add 2 ml NED reagent to standards, blank, and sample. Wait 10 min for color development.
6. Read on absorbance on spectrophotometer at 543 nm. Prepare calibration curve and calculate concentration.

Federal Limits

Primary MCL: 1 mg/l.

Reference

American Public Health Association, American Water Works Association, and Water Environment Federation, *Standard Methods for the Examination of Water and Wastewater*, 20th ed., American Public Health Association, Washington, D.C., 1998, 4-112.

Nitrate Nitrogen

Nitrate (NO_3^-) is the stable, most oxidized form of nitrogen found in natural waters. It is normally present in only small concentrations (<1 mg/l). When larger concentrations occur in ground or surface waters, the most common cause is runoff or seepage from excessive agricultural fertilizer applications. Nitrate in potable waters is limited (primary MCL10 mg/l) because it is believed to contribute to the occurrence of methylemoglobinemia (blue baby) in infants.

Nitrifying bacteria in natural waters will oxidize ammonia to nitrite, then to nitrate. These bacteria can be found in clean surface waters and distribution piping.

Nitrification

$$NH_3 \xrightarrow[\text{Nitrosomonas}]{} NO_2^- \xrightarrow[\text{Nitrobacter}]{} NO_3^-$$

Treatment Plant Significance

Conventional potable water treatment plants cannot remove nitrate. High concentrations must be prevented by controlling the input at the source.

Analysis: Cadmium Reduction Method for Nitrate Nitrogen

Nitrate is reduced to nitrite by passing the sample through a column packed with activated cadmium. The sample is then measured quantitatively for nitrite.

Quality Control

- Assure accuracy of standards.
- Compare results of this method with results of nitrate electrode method.
- Test should be done within 48 h of sampling. The nitrate present is fairly stable, but with standing, the ammonia and nitrite in the sample will oxidize, yielding false high results. If septic conditions occur, nitrate will be reduced to nitrogen gas.
- Store nitrate electrode in 100 mg/l nitrate standard solution. Replace endings monthly; replace electrode after 1 year.
- Nitrate standards have a shelf life of 2 weeks.

Apparatus

- Reduction column
- Spectrophotometer

Reagents

Nitrate-free distilled water

Pass distilled water through deionizer column appropriate for removing nitrate and store in tightly stoppered glass container. Use for preparing reagents, standards, and for making dilutions.

Copper cadmium granules

Reduction material

Wash 25 g Cd granules (40 to 60 mesh) with 6 *N* HCl and rinse with water. Swirl Cd with 100 ml 2% $CuSO_4$ solution for 5 min, decant, and repeat until brown precipitate begins to develop. Flush with water to remove all precipitate.

NED reagent

N-(1-naphthyl)-ethylenediamine dihydrochloride

Produces the color

Add 100 ml 85% phosphoric acid and 10 g sulfanilamide to 80 ml water. When dissolved, add 1 g NED; mix, and dilute to 1 l with water. Store in dark bottle. Reagent is stable for 1 month.

Ammonium chloride–EDTA solution

Activates reduction column

Dissolve 13 g NH_4Cl and 1.7 g EDTA in 900 ml water. Adjust to pH 8.5 with concentrated NH_4OH and dilute to 1 l.

Dilute ammonium chloride–EDTA solution

To rinse reduction column

Dilute 300 ml NH_4Cl-EDTA to 500 ml with water.

Copper sulfate solution

$CuSO_4 \cdot 5H_2O$, 2%

To prepare Cd granules

Dissolve 20 g $CuSO_4 \cdot 5H_2O$ in 500 ml water and dilute to 1 l.

Stock nitrate solution

KNO_3

The standard

Dry potassium nitrate (KNO_3) at 105°C for 24 h. Dissolve 0.7218 g in water and dilute to 1 l (100 mg/l NO_3-N). Prepare fresh every 2 weeks. Use this solution to prepare working standards.

Stock nitrite solution

$NaNO_2$, 250 mg/l, 0.0179 *M*

The standard

Dissolve 1.232 g $NaNO_2$ in water and dilute to 1 l. Must be standardized. Prepare working standards from this solution. Make fresh daily.

Standard potassium permanganate

$KMnO_4$, 0.05 *M*

Used to standardize the nitrite stock solution

Dissolve 8 g $KMnO_4$ in 1 l water. Keep in brown bottle and age for at least 1 week. Must be standardized.

Sodium oxalate

$Na_2C_2O_4$, 0.05 *N*

Primary standard, to standardize the permanganate

Dissolve 3.35 g $Na_2C_2O_4$ in water and dilute to 1 l.

Procedure

1. Standardize potassium permanganate:

   Dissolve 200 mg sodium oxalate in 100 ml water.

   Add 10 ml 1+1 H_2SO_4 and heat to 95°C. Keep heat above 85°C throughout the entire procedure.

   Titrate permanganate solution into this until faint pink color appears and persists for 1 min.

   Calculate: $$\text{Molarity KMnO}_4 = \frac{\text{g Na}_2\text{C}_2\text{O}_4}{\text{ml titrated} \times 0.335}$$

   Adjust to 0.05 *M*.

2. Standardize nitrite stock solution:

   Add, in this order, 50 ml standard 0.05 *M* $KMnO_4$, 5 ml concentrated H_2SO_4, and 50 ml stock nitrite solution.

   Warm to 80°C on hot plate.

   Titrate $Na_2C_2O_4$ into this solution until permanganate color is just barely discharged.

   Calculate molarity of stock solution.

3. Prepare working standards 0.2 to 1.0 mg/l, or expected range. Prepare 50 ml of each.

4. Prepare reduction column:

   Insert glass wool plug into bottom of column and fill with water.

   Add enough Cu-Cd granules to fill to 18.5 cm.

   Wash column with 200 ml diluted NH_4Cl-EDTA solution.

   Activate column by passing it through 100 ml of a solution composed of 25% 1 mg/l NO_3-N standard and 75% NH_4Cl-EDTA solution (7 to 10 ml/min).

5. Adjust sample pH to between 7 and 9.

6. Reduce sample:

   To 25-ml sample add 75 ml NH_4Cl-EDTA and mix.

   Pour through column (7 to 10 ml/min). Discard the first 25 ml.

7. To reduced sample (and nitrite standards) immediately add 2 ml NED reagent, mix, and wait 10 min for color development.

8. Read absorbance on spectrophotometer at 543 nm. Prepare calibration curve and calculate concentration.
9. Prepare nitrate standards and run these through column. Test reduced standards for NO_2-N to verify reduction column efficiency.

Analysis: Nitrate Electrode Method for Nitrate Nitrogen

The nitrate electrode has an ion exchange material contained in a solid plastic membrane. It acts as a small ion exchanger which exchanges the nitrate ion with hydrogen ions across the membrane, creating the voltage potential. A separate reference electrode is needed for this measurement (there will be two probes in the sample solution). The reference electrode is a typical pH reference electrode, but one with a double junction, a separate chamber that allows the user to choose a different filling solution.

Quality Control

- Chloride and bicarbonate ions in substantial amounts interfere with the electrode reading. ionic strength adjustment (ISA) buffer, an acid, is added to the sample to prevent bicarbonate interference by changing bicarbonate to carbonic acid.
- Assure accuracy of standards.
- Test should be done within 48 h of sampling. The nitrate present is fairly stable, but with standing, the ammonia and nitrite in the sample will oxidize, yielding false high results. If septic conditions occur, nitrate will be reduced to nitrogen gas.
- Store nitrate electrode in 100 mg/l nitrate standard solution. Replace endings monthly and replace electrode after 1 year.
- Nitrate standards have a shelf life of 2 weeks.

Apparatus

- pH meter
- Double junction reference electrode
- Nitrate selective electrode

Reagents

Nitrate-free distilled water

Pass distilled water through deionizer column appropriate for removing nitrate and store in tightly stoppered glass container. Use for preparing reagents, standards, and for making dilutions.

Stock nitrate solution

KNO_3, 100 mg/l NO_3-N
The standard
Dry potassium nitrate (KNO_3) at 105°C for 24 h. Dissolve 0.7218 g in water and dilute to 1 l. (100 mg/l NO_3-N). Prepare fresh every 2 weeks. Use this solution to prepare working standards.

Buffer solution

Prevents bicarbonate interference
Dissolve 17.32 g $Al_2(SO_4)_3 \cdot 18H_2O$, 3.43 g Ag_2SO_4, 1.28 g H_3BO_3, and 2.52 g sulfamic acid (H_2NSO_3H) in 800 ml water. Adjust to pH 3 with 0.1 *N* NaOH. Dilute to 1 l and store in dark glass bottle.

Sodium hydroxide

NaOH, 0.1 *N*
Buffer preparation
Dissolve 4 g NaOH in water and dilute to 1 l.

Procedure

1. Place 20 ml of standard or sample in a small beaker and add 20 ml buffer. Mix. Use standards of 0.1, 10, and 100 mg/l.
2. Immerse electrodes, read millivolts, and plot on semilog graph paper.

Federal Limits

10 mg/l (Primary MCL, NO_3-N)
1 mg/l (Primary MCL, NO_2-N)

Reference

American Public Health Association, American Water Works Association, and Water Environment Federation, *Standard Methods for the Examination of Water and Wastewater*, 20th ed., American Public Health Association, Washington, D.C., 1998, pp. 4-116, 4-117.

Chapter 25

Phosphorus

Contents

Phosphorus is a vital nutrient for all living things. Cellular phosphate compounds trap energy generated from food consumed and transfer it to activities that demand it: locomotion, reproduction, and growth. Without the phosphorus to build these energy compounds, cell life cannot exist.

Excessive phosphorus from runoff or wastewater discharges in natural water bodies, however, stimulates bacterial and algal growth. Massive blooms may occur with resultant deposition of excessive solids eutrophication and eventually fill up the water body, shortening its life. Phosphorus occurs naturally, almost solely as phosphates. Most phosphates are dissolved, but some are in combination with suspended particles in the water and may contribute to turbidity.

Phosphates occur in three forms:

Orthophosphate: Simple phosphates, or reactive phosphate, i.e., Na_3PO_4, sodium phosphate (tribasic), NaH_2PO_4, and sodium phosphate (monobasic). Orthophosphate is the only form of phosphate that can be directly tested for in the laboratory, and is the form that bacteria use directly for metabolic processes.

Polyphosphate: Acid hydrolyzable phosphate, i.e., $Na(PO_3)_x$ sodium hexametaphosphate (calgon). Polyphosphates come from detergents and water additives. They can be converted to orthophosphate by acid addition and by boiling the sample.

Organic Phosphate: In most natural waters this is a very small concentration. It can be converted to orthophosphate by digestion with an oxidizing agent under strong acid conditions. This digestion also converts polyphosphate to orthophosphate. All phosphates will gradually hydrolyze in natural waters to the ortho form.

Total Phosphorus: This is the total amount of phosphorus in the sample after all forms have been converted to orthophosphate. It is total phosphate that is most often tested for.

Treatment Plant Significance

Phosphorus is normally very low (<1 mg/l) in clean potable water sources, and is not regulated by the federal government in drinking waters. If necessary, removal can be achieved by the addition of coagulant metal salts, frequently combined with an anionic polymer. The coagulants react with water to produce floc particles that adsorb turbidity in the water and settle it out. In this manner, phosphorus components, which are associated with solid particles, are removed. Some dissolved phosphorus will also be converted to particulate form and removed. It is not common to see treatment specifically for phosphorus removal in potable water systems. Some do monitor phosphorus, however, particularly if the source water is of high organic nature, if the treatment plant adds phosphorus for corrosion or scale control, or if there is a problem with regrowth of bacteria in distribution system piping.

Analysis: Total Phosphorus

A. Persulfate Digestion Method for Conversion of Organic Phosphorus and Polyphosphorus to Orthophosphorus

Persulfate digestion is a safe, simple procedure for conversion of more complex forms of phosphorus to orthophosphate.

Quality Control

- Glassware for phosphorus testing must be acid washed to eliminate any soap residue that may contain phosphorus. Wash in hot water with phosphorus-free soap, rinse twice with 10% HCl solution, and then with distilled or deionized water.

- Samples for phosphorus testing should not be kept more than 2 days.
- Samples should be acidified and kept refrigerated at 4°C, if they must be stored at all.

Apparatus

- Acid-washed glassware
- Hot plate

Reagents

Sulfuric acid solution

H_2SO_4
Catalyzes the digestion
Add 300 ml concentrated sulfuric acid to 600 ml distilled water and dilute to 1 l.

Sodium hydroxide

NaOH, 1 *N*
To neutralize sample after digestion
Dissolve 40 g NaOH in distilled water and dilute to 1 l.

Ammonium persulfate

$(NH_4)_2S_2O_8$
Digestion reagent
Dry, solid

Procedure

1. Adjust sample to phenolphthalein endpoint.
2. Add 1 ml sulfuric acid solution and 0.4 g ammonium persulfate to 50-ml sample.
3. Boil gently for 40 min, or until volume is down to 10 ml.
4. Cool and dilute to 30 ml with distilled water. Neutralize to phenolphthalien endpoint with NaOH.
5. Make up to 50 ml with distilled water. If precipitate forms, ignore it. It will redissolve in the acid conditions of the orthophosphorus test.

B. Ascorbic Acid Method for Determining Orthophosphate

Ammonium molybdate and potassium antimonyl tartrate react under acid conditions with orthophosphate to produce a blue phosphomolybdic acid complex. Compared to orthophosphate standards and graphed, concentration can be determined.

Absorbance is read on the spectrophotometer at a wavelength of 880 nm, out of the visible light range. Older spectrophotometers that cannot achieve the higher wavelengths may find a good peak at 650 nm (also accepted by EPA). The ascorbic acid method is the most accurate method for waters with small concentrations of phosphorus (under 1 ppm). If sample is expected to have a higher concentration of phosphate, dilute it.

Quality Control

- Phosphate standards must be weighed out carefully and exactly.
- Standards should be carried through the entire digestion procedure along with the samples.
- Allow the same amount of time for color development for all standards and samples.
- Compare sample concentration calculated from slope against sample concentration visually observed on graph.

Apparatus

- Acid-washed glassware
- Spectrophotometer

Reagents

Sulfuric acid

H_2SO_4, 5 *N*
Catalyzes reaction
Dilute 70 ml concentrated H_2SO_4 to 500 ml with distilled water.

Potassium antimonyl tartrate solution

$K(SBO)C_4H_4O_6 \cdot 1/2\ H_2O$
Reacts to form complex which can be reduced by ascorbic acid to produce blue color in presence of phosphorus
Dissolve 1.3715 g potassium antimonyl tartrate in 400 ml distilled water and dilute to 500 ml.

Ammonium molybdate solution

$(NH_4)_6Mo_7O_{24} \cdot 4H_2O$
Reacts to form complex which can be reduced by ascorbic acid to produce blue color in presence of phosphorus. Dissolve 20 g ammonium molybdate in 500 ml distilled water.

Ascorbic acid

1 *M*

Reduces the above reagents to produce blue color in the presence of phosphorus

Dissolve 1.76 g ascorbic acid in 100 ml distilled water. Reagent will remain stable for 1 week.

Combined reagent

Produces the color

Mix in this order: 50 ml 5 *N* H_2SO_4, 5 ml potassium antimonyl tartrate solution, 15 ml ammonium molybdate solution, and 30 ml ascorbic acid solution. Mix well after addition of each. Reagent will remain stable for 4 h.

Stock phosphate solution

KH_2PO_4

The standard

Dissolve 219.5 mg anhydrous KH_2PO_4 in distilled water and dilute to 1 l.

Procedure

1. Acid wash all glassware with hot diluted HCl.
2. If turbidity exists in raw sample, filter through a 0.45-μm membrane filter. If it is present in digested sample, ignore it; acidic reagent should dissolve it.
3. If expected sample concentration is above 2 mg/l, dilute sample.
4. Adjust sample pH to phenolphthalein endpoint.
5. Prepare working standards to cover range.
6. Add 8 ml combined reagent to standards, reagent blank, and sample. Mix and wait 10 min for color development.
7. Read absorbance on spectrophotometer (880 nm). Prepare calibration curve and calculate sample concentration.

Federal Limits

None.

Reference

American Public Health Association, American Water Works Association, and Water Environment Federation, *Standard Methods for the Examination of Water and Wastewater*, 20th ed., American Public Health Association, Washington, D.C., 1998, p. 4-146.

Chapter 26

Atomic Absorption Spectrometry

Contents

Metals are those elements that easily lose electrons to form positive ions. They form the cation portion of salts and are necessary as trace elements required for cell metabolism. The method described below is in the category of flame emission photometry, and is an EPA-approved method that the water utility industry uses to test for metals.

When the atoms of a metal are heated enough to excite the electrons, light is given off at certain wavelengths that are characteristic of that metal only. An instrument invented in the 1800s, the spectroscope, identifies these wavelengths and the metal corresponding to them. Many of the less common metals were first discovered in this way. Several wavelengths of light are emitted for each metal: some in the visible light range, some in ultra violet, and some in infra-red. Each metal has its own characteristic series of "spectral lines." Light at one of these wavelengths, however, will be emitted more strongly than at the others.

When a water sample containing a particular metal is vaporized in a colorless flame, that flame will absorb the wavelengths of light characteristic for that metal and give off other colors from remaining components of the sample. Therefore, a flame color is seen which is a mixture of different kinds of light, but the ones that are characteristic for that metal are missing. They have been absorbed by the flame. The flame color then passes through a prism that breaks up the light into its component colors. They will all show, except the missing ones. These will show up as dark spaces, or "dark spectral lines." The **atomic absorption spectrometer** expands upon this idea. The sample is vaporized in an acetylene flame; the light from the flame passes through a prism which breaks it up. A chosen wavelength is then allowed to pass through a slit in an opaque plate. The wavelength chosen is that which will be most strongly absorbed by the flame for the metal it is to identify. Light coming through the slit is picked up by a phototube and registers on an analog scale, or is changed to a digital readout. The wavelength that is missing because of the flame absorption is added by the instrument. A special light bulb (hollow cathode tube) is used, which emits light only at the wavelength characteristic for that particular metal. This adds the metals color into the flame.

If the flame has spent its energy absorbing light from the metal in the sample, it is less likely to absorb the light from the light bulb, and that light will register on the phototube. If there were little or none of that metal in the sample, the flame would absorb the light from the light bulb, and much less of this light would register on the phototube. The result is a stronger or weaker reading for the particular metal being analyzed. When compared with standards, concentration can be calculated.

The presence of other elements in the sample will not interfere. The instrument looks just at the one wavelength characteristic of the metal being analyzed. The actual flame color seen by the chemist is not necessarily significant, and may be wavelengths from other metals in the sample. sodium yellow is most often seen in concentrated samples.

Except for the vaporizing of the sample in the flame and the use of the hollow cathode tube, this method is similar to visible light spectrophotometry. No color reagents are needed and the flame and hollow cathode tube perform the sensing of the proper color.

Treatment Plant Significance

High concentrations of metals may exist in water due to industrial waste discharges from manufacturing and metal finishing in surface waters. Copper and lead in drinking water are the result of distribution piping and solder corrosion. Metals in clean groundwaters are of geologic origin. Metals are also monitored in wastewater effluents, and in sludges applied to land for protection of the groundwater. Some ordinary metal concentrations are present in wastewaters chiefly because they existed in the source water.

The most economical method to remove most metals is ferric sulfate coagulation/settling done at pHs of 7 to 9. Ion exchange, reverse osmosis, electrodialysis, and softening processes may be used for certain metals. Conventional filtration plants do not remove metals, nor do most wastewater treatment processes.

Listed below are metals of interest with their health effects at high concentrations in water:

Aluminum

Source: Geological; chemical additive for water treatment

Effect: Under study

Antimony

Source: Geological; manufacture of ceramics, glass, and pesticides

Effect: Decreases growth and longevity

Arsenic

Source: Geological

Effect: Carcinogenic; causes cyanosis

Beryllium

Source: Geological; manufacture of high heat materials

Effect: Bone and lung damage

Barium

Source: Paint and paper manufacture

Effect: Circulatory system damage and nerve block

Cadmium

Source: Geological, mining, smelting, and metal plating

Effect: Kidney damage and hypertension

Calcium

Source: Geological

Effect: No adverse effect on humans or water hardness

Chromium

Source: Metal plating

Effect: Liver, kidney damage, and skin disorders

Copper

Source: Geological; corrosion of interior piping

Effect: Gastric distress and Wilson's disease

Iron

Source: Geological

Effect: No adverse effect on humans

Lead

Source: Lead water pipe and lead-based pipe solder, lead-based paints, and dust.

Effect: Central and peripheral nervous system damage, kidney damage, increased blood pressure, hearing impairment, interferes with red blood cell synthesis, anemia, possible carcinogen, premature birth, and reduced birth weight; affects infants and small children most seriously

Magnesium

Source: Geological

Effect: No adverse effect on humans or water hardness

Manganese

Source: Geological

Effect: No adverse effect on humans

Mercury

Source: Geological; paint, paper, fungicide manufacture, and burning of fossil fuels

Effect: Inflammation of mouth and gums, swelling of salivary glands, tooth loss; central nervous system and kidney damage

Nickel

Source: Geological; electroplating and manufacture of ceramics and batteries

Effect: Heart and liver damage

Potassium

Source: Geological

Effect: No adverse effect on humans

Selenium

Source: Geological and mining; manufacture of electrical components

Effect: Gastrointestinal effects; red staining of fingers, teeth, and hair; weakness, depression, and irritation of nose and throat

Sodium

Source: Geological

Effect: May adversely affect people with heart disease or hypertension; the American Heart Association recommends 20 mg/l limit in water supplies

Silver

Source: Geological; mining

Effect: Gray discoloration of skin, eyes, and mucous membranes

Thallium

Source: Geological; electronics, alloys, and glass manufacture

Effect: Kidney, liver, and brain damage

Zinc

Source: Mining; electroplating and pipe corrosion

Effect: Low toxicity

Analysis: Atomic Absorption Method for Metals

Most metals analysis is performed by atomic absorption spectrometry. Metals may require pretreatment for dissolution, concentration, and separation from organics. For a few, it is necessary to dilute the sample.

Quality Control

- Standards are prepared "as the metal." See *Standard Methods* for preparation.
- Calibrate instrument according to directions. Adjust to obtain largest peak for the metal desired.

Apparatus

- Atomic absorption spectrometer

Reagents

None, except those needed for metal pretreatment

Procedure

1. Prepare standards according to instrument guidelines. The standard concentrations are limited to those that will produce a linear slope.
2. Concentrate or dilute sample to read within standards.
3. Aspirate samples and standards.
4. Graphing and calculation of concentration may be manual or done automatically by the instrument.

Federal Limits

0.01 mg/l (Primary MCL, As)
1.0 mg/l (Primary MCL, Ba)
0.004 mg/l (Primary MCL, Be)
0.005 mg/l (Primary MCL, Cd)
0.1 mg/l (Primary MCL, Cr)
0.002 mg/l (Primary MCL, Hg)
0.01 mg/l (Primary MCL, Ni)
0.05 mg/l (Primary MCL, Se)
0.006 mg/l (Primary MCL, Sb)
0.005 mg/l (Primary MCL, Tl)
0.1 mg/l (Secondary MCL, Ag)
0.2 mg/l (Secondary MCL, Al)
1.0 mg/l (Secondary MCL, Cu)
0.3 mg/l (Secondary MCL, Fe)
0.05 mg/l (Secondary MCL, Mn)
5 mg/l (Secondary MCL, Zn)

Reference

American Public Health Association, American Water Works Association, and Water Environment Federation, *Standard Methods for the Examination of Water and Wastewater*, 20th ed., American Public Health Association, Washington, D.C., 1998, p. 3-12. See index for individual metals.

Appendix I

Practice Problems

General

1. An empty container weighs 2 kg. Filled with water it weighs 23 lb. What is the volume of the container (liters)?

   Ans. 8.44 l

2. Change to concentration in mg/l:
   a. 0.2 lb/l
   b. 1 lb/gal
   c. 0.006%
   d. 12 g/MG

   Ans. a. 90,800 mg/l
   b. 119,947 mg/l
   c. 60 mg/l
   d. 0.003 mg/l

3. Change 65 mg/l to units of % concentration.

   Ans. 0.0065%

4. One g of salt is dissolved in 30 gal of water. What is the concentration in mg/l?

   Ans. 8.8 mg/l

5. How many grams is 3.5 lb. equivalent to?

   Ans. 1589 grams

6. What is the normality of a 500-mg/l solution of phosphoric acid?

   Ans. 0.015 *N*

7. A solution is prepared at a concentration of 400 µg/l. What is the concentration in mg/l?

   Ans. 0.4 mg/l

8. A 0.06 *N* solution of sulfuric acid is what concentration in mg/l?

Ans. 2940 mg/l

9. A 0.05 *N* solution of sulfuric acid has been prepared. Find:
   a. The molarity
   b. The concentration in mg/l
   c. The % concentration
   d. How much of it is needed to prepare 50 ml of a 0.012 *N* solution?

Ans. a. 0.025 *M*
b. 2450 mg/l
c. 0.2450%
d. 12 ml

10. How much of a 0.5 *N* sulfuric acid solution is needed to prepare 100 ml of a 0.125 *N* solution?

Ans. 25 ml

11. What is the mg/l concentration of a 1 *N* HCl solution?

Ans. 36,000 mg/l

12. What is the molarity of a 450-ml HCl solution that contains 20 g of HCl?

Ans. 1.24 *M*

13. How many ml of 0.3 *N* HCl will neutralize 50 ml of 0.0125 *N* NaOH?

Ans. 2.1 ml

14. 20 ml of 2 *N* HCl is mixed with 480 ml of 0.5 *N* HCl. What is the concentration of the final mixture?

Ans. 0.56 *N*

15. If 50 ml of a 0.5 *N* NaOH solution are required to neutralize 25 ml of acid solution, what is the normality of the acid?

Ans. 1 *N*

16. Concentrated hydrochloric acid has a specific gravity of 1.18. The solution, as purchased, is 37% acid. How many ml of this acid are needed to prepare 500 ml of a 0.2 *N* HCl solution?

Ans. 8.24 ml

17. 20 ml of element X (approximately 0.01 *N*) should neutralize 20 ml of element Y (exactly 0.01 *N*) in a titration. However, it actually takes 26 ml of element X to complete the neutralization. What is the normality of element X?

Ans. 0.008 *N*

18. 38 mg of chemical is dissolved in 30 ml water. What is the concentration in mg/l?

Ans. 1267 mg/l

19. 30 ml of a 0.01 *N* solution of KOH is added to 50 ml of a 1400 mg/l solution of KOH. What is the normality of the mixed solution?

Ans. 0.0194 *N*

20. 5 lb of a contaminant per million gallons of water, is what concentration in mg/l?

Ans. 0.6 mg/l

Alkalinity/Hardness/Softening

1. Testing for alkalinity with 0.02 *N* H_2SO_4, 25 ml was used for the titration into a 100-ml sample. What was the total alkalinity, as $CaCO_3$?

Ans. 250 mg/l

2. In the same 100-ml sample above, if 0.1 *N* H_2SO_4 were used to do the titration, how many ml would be needed?

Ans. 50 ml

3. In the same 100-ml sample, if 0.02 *N* hydrochloric acid were used, how many ml would be needed?

Ans. 25 ml

4. If the sample were 25 ml in size, and 0.02 *N* H_2SO_4 were used, what would be the total alkalinity?

Ans. 1000 mg/l

5. Calculate the concentrations of the total alkalinity and of the different component types of alkalinity, as $CaCO_3$, in these three 100-ml samples (0.02 *N* H_2SO_4 used as titrant).

Sample	pH	ml to Pheno. Endpoint	Total ml titrated
1.	10.8	10.3	26.0
2.	7.7	0.0	8.8
3.	10.2	13.1	26.2

Ans. 1. Tot. Alk.: 260 mg/l
Hydroxide: 0
Carbonate: 206 mg/l
Bicarbonate: 54 mg/l

2. Tot. Alk.: 88 mg/l
Hydroxide: 0
Carbonate: 0
Bicarbonate: 88 mg/l

3. Tot. Alk.: 262 mg/l
Hydroxide: 0
Carbonate: 262 mg/l
Bicarbonate: 0

6. Lab data: pH is 7.9, total hardness is 250 mg/l, as $CaCO_3$. Bicarbonate alkalinity is 285 mg/l, as $CaCO_3$. What is the noncarbonate hardness?

Ans. 0

7. Alkalinity is measured in a new 100-ml water sample. The pH is 10.4, phenolphthalein is added, and it takes 13 ml to reach the color change. Bromocresol green is added and it takes 4 more ml to reach the total alkalinity endpoint. If the titrant is 0.02 *N* H_2SO_4, how much of each type of alkalinity is in the sample (as $CaCO_3$)?

Ans. Hydrox.: 90 mg/l
Carbonate: 80 mg/l
Bicarbonate: 0

8. A 100-ml sample (pH 8.6) titrated with 0.02 *N* H_2SO_4 yielded a total alkalinity of 160 mg/l, as $CaCO_3$, Of this, 20 mg/l was carbonate alkalinity, and 140 mg/l was bicarbonate alkalinity. The sample had a total hardness of 150 mg/l (all as $CaCO_3$).

a. How many ml was titrated to the phenolphthalein endpoint?

b. What was the hydroxide alkalinity?

c. What was the noncarbonate hardness?

Ans. a. 1 ml
b. 0
c. 0

9. Convert to concentrations expressed as $CaCO_3$:

260 mg/l Ca^{2+}, as Ca^{2+} — Ans. 650 mg/l

80 mg/l $(CO_3)^{2-}$, as $(CO_3)^-$ — Ans. 133.3 mg/l

220 mg/l $(HCO_3)^-$, as $(HCO_3)^-$ — Ans. 180.3 mg/l

10. A 500-ml sample of water was sent to a private lab for testing hardness and alkalinity. Lab results were reported as:

Total alkalinity: 304 mg/l, as $CaCO_3$
Total hardness: 206 mg/l, as $CaCO_3$
$(OH)^-$ 24 mg/l, as $CaCO_3$
$(CO_3)^{2-}$ 80 mg/l, as $CaCO_3$
$(HCO_3)^-$ 200 mg/l, as $(HCO_3)^-$
Ca^{2+} 20 mg/l, As Ca^{2+}
Mg^{2+} 38 mg/l, as Mg^{2+}

There is an error in these results. What is it?

Ans. Cannot have all three forms of alkalinity in one water sample.

11. Determine lime and soda ash dosage needed to soften this water. Chemicals are purchased 100% pure:

pH 7.8
Total alkalinity 340 mg/l, as $CaCO_3$
Total hardness 360 mg/l, as $CaCO_3$
Ca^{2+} hardness 280 mg/l, as $CaCO_3$

CO_2 38 mg/l, as $CaCO_3$

Excess lime desired: 30 mg/l, as $CaCO_3$ Ans. lime: 273 mg/l

soda ash: 21 mg/l

12. The municipal water softening plant adds 280 mg/l lime to its 14 MGD flow. How many lb of lime do they use each year?

Ans. 11,932,872 lb/yr

Chemical Dosage/Jar Testing

1. A 1% aluminum sulphate stock solution to be used for jar testing is what concentration in mg/l?

   Ans. 10,000 mg/l

2. An industrial alum solution (17% active aluminum) is to be used for jar testing. How many ml of this are to be added to the 1-l stock solution flask to achieve a concentration of 0.5% active aluminum?

   Ans. 29.4 ml

3. If 0.2 lb of dry coagulant is mixed with each gal of water to prepare a stock solution for feeding into the plant flow, what is the percent concentration of the coagulant in the stock solution?

   Ans. 23,989 mg/l

4. For a jar test, 0.05 lb coagulant chemical is dissolved in a l of water. 3 ml of this stock is added to a one l jar for flocculation. What is the concentration of the coagulant in the jar?

   Ans. 68 mg/l

5. In preparing a jar test, how many ml of a 2% coagulant stock solution should be added to a 1000-ml jar to achieve a concentration of 25 mg/l?

   Ans. 1.25 ml

6. If 1 ml of a 2.3% alum solution is added to a 500-ml jar, what is the alum concentration in the jar?

   Ans. 46 mg/l

7. For jar testing, a stock solution of alum was prepared by dissolving 1 g of $Al_2(SO_4)_3$ in 1 l of water. How many ml of this solution should be added to a 1000-ml jar to achieve a concentration of 6 mg/l aluminum ion?

   Ans. 38 ml

8. 2.6 lb alum ($Al_2(SO_4)_3{\cdot}6H_2O$) is dissolved in 100 gal of water. What is the aluminum ion concentration (mg/l)?

   Ans. 374 mg/l

9. By jar testing, it was determined that 3 ml of a 19 g/l alum solution in a 1000-ml sample would be the correct dosage. How many lb per day of the dry chemical must be added to a 700 gal/min flow to simulate this jar test dosage?

Ans. 475.4 lb/day

Chlorine Residual/Demand

1. Commercial liquid chlorine bleach is 5.25% hypochlorite ion (active chlorine). What is this concentration in mg/l?

Ans. 52,500 mg/l

2. A jug of liquid chlorine bleach (5.25% active chlorine) is used to prepare a standard for chlorine residual analysis. How many ml is needed to prepare a 100-mg/l concentration in a 500-ml flask?

Ans. 0.95 ml

3. What is the percentage of active chlorine in a tank of chlorine gas?

Ans. 100%

4. What is the chlorine demand of a water whose chlorine residual is 0.5 mg/l, if a chlorine dose of 14 lb/day is being added to the 500-gpm flow?

Ans. 1.8 mg/l

5. 20 ml of a 100 mg/l chlorine standard is added to 1 liter of process water. After the appropriate contact time, the chlorine residual is read at 0.7 mg/l. What was the chlorine demand of this water?

Ans. 1.3 mg/l

6. How many ml of sodium hypochlorite (12% available chlorine) must be added to a liter of water to prepare a 1000 mg/l available chlorine solution?

Ans. 8.3 ml

7. If 15 lb of chlorine gas is dissolved into 1500 gal of pure water, what is the chlorine concentration in mg/l?

Ans. 1199 mg/l

8. How many lb of free available chlorine is there in 20 lb of $Ca(OCl)_2$?

Ans. 14.2 lb

9. How many grams of calcium hypochlorite must be dissolved in 1 liter of water to prepare a 0.1% available chlorine solution?

Ans. 1.4 g

10. The chlorine demand of a water is 8 mg/l. How many gallons of 15% sodium hypochlorite weighing 9.3 lb/gal must be added each day to the 6 MGD flow to achieve a chlorine residual of 1 mg/l?

Ans. 323 gal

Chloride

1. From *Standard Methods*, the directions to prepare a solution read, "Dissolve 1.64 g of the chloride standard in distilled water and dilute to 1000 ml: 1 ml = 1 mg Cl^-."

   What is the mg/l concentration of this solution?

   Ans. 1000 mg/l

2. How much NaCl would be needed to prepare 1 liter of 1000 mg/l chloride stock solution?

   Ans. 1667 mg

3. If 2 g NaCl is added to 1000 ml water, what is the percent concentration?

   Ans. 0.2%

4. A solution of 500 ml of a 5% NaCl was boiled to dryness and left a white residue on the container. How many grams of this residue was chloride?

   Ans. 15 g

5. Sea water is 5% salt (sodium chloride). How many pounds of chloride are in 1 million gal of sea water?

   Ans. 250,200 lb

6. A 2% solution of NaCl, as Cl^- is what concentration in mg/l, as NaCl?

   Ans. 33,333 mg/l

7. A 500-mg/l chloride standard is used to spike a 100-ml water sample with 5 mg/l chloride. How many ml of the standard should be added to the sample to achieve this spike?

   Ans. 0.1 ml

8. A 1 MGD seacoast community is forced to desalinate the water supply because the chloride concentration is 700 ppm. What percent chloride is this?

   Ans. 0.07%

9. A conventional filtration plant treats a flow of 6 MGD. If the chloride concentration of the influent water is 25 mg/l, what is the chloride concentration of the filtered water?

   Ans. 25 mg/l

10. If 100 ml of water with a chlorine demand of 7 mg/l was poured into a flask that contained 500 ml of a 35 mg/l sodium chloride solution, what was the chlorine residual in the flask after mixing?

Ans. 0

Fluoride/Sulphate

1. How many ml of water are needed to reduce 500 ml of a 0.1% solution of NaF to 0.01 *N*?

Ans. 690.5 ml

2. To prepare 1 l of 500 mg/l fluoride ion, NaF is used. How many mg of the chemical has been dissolved in the liter flask?

Ans. 1106 mg

3. In preparing 100 ml of a working standard that will contain 0.4 mg/l fluoride ion, a 0.01% stock solution of sodium fluoride is used. How many ml of the stock is needed to prepare the working standard?

Ans. 0.9 ml

4. If 2 ml of a 0.02% fluoride ion solution is added to a 250-ml flask of water that has a natural fluoride content of 2.8 mg/l, what is the final fluoride concentration in the flask?

Ans. 4.4 mg/l

5. In preparing 100 ml of a working standard that will contain 1.8 mg/l fluoride ion, a 0.2% stock solution of hydrofluosilicic acid (the stock solution is 5% acid) is used. How many ml of the stock is needed to prepare the working standard?

Ans. 2.3 ml

6. Sodium silicofluoride is used to prepare 250 ml of a 500 mg/l solution of fluoride ion. How many mg is needed?

Ans. 205 mg

7. A sulfate standard is prepared at 100 mg/l, as sulfate. The chemical used is sodium sulfate. How much of the chemical must be weighed out to prepare a liter of this standard?

Ans. 147 mg/l

8. A 0.02 *N* sulfuric acid solution was available in the lab from performing the total alkalinity test. If it were to be used as a standard for the sulfate test, what mg/l concentration would that be?

Ans. 975 mg/l

9. A 1 *N* solution of sodium sulfate is prepared as a standard for a sulfate test. How many mg/l is this solution, as sulfate?

Ans. 48,300 mg/l

10. A water high in sulfate was diluted ten times and then measured colorimetrically. If the absorbance reading was 0.2 and the standard slope was 38, what is the sulfate concentration of the original sample?

Ans. 76 mg/l

Solids

1. Calculate mg/l total suspended solids and % volatile in the following samples:

a.	Tare wt	23.0788 g
	Dry wt	23.0851 g
	Ash wt	23.0811 g
	Sample size	50 ml
b.	Tare wt	23.9123 g
	Dry wt	23.9138 g
	Ash wt	23.9127 g
	Sample size	100 ml

Ans. a. TSS 126 mg/l
64% volatile
b. TSS 15 mg/l
73% volatile

2. A water sample has 1 g/gal total solids. From the solids test, the TSS concentration is found to be 200 mg/l. What is the concentration of the dissolved solids in this sample?

Ans. 64 mg/l

3. A 200-ml sample is filtered for total dissolved solids analysis. Dry solids weight is 0.15 g. What is the mg/l concentration of the TDS in this sample?

Ans. 750 mg

4. For a total solids test, the 70-ml sample is dried and weighed and yields a dry solids content of 0.03 g. What is the mg/l concentration of total solids?

Ans. 429 mg/l

5. On a clean water sample that is high in total solids, or total dissolved solids, what are the most likely sources of these solids?

Ans. High alkalinity and hardness

6. A lime-softening sludge has a 2% solids concentration. If the process flow is 50 MGD and the sludge flow is 4% of the process flow, how many lb per day dry solids must this utility dispose of?

Ans. 333,600 lb/day

7. If the raw water has 45 mg/l suspended solids and 90% of the suspended solids is removed by filtration, what is the suspended solids concentration of the filtered water?

Ans. 4.5 mg/l

8. If backwashing is done at 37,500 gpm for 20 min twice a week, and the backwash water is 0.5% solids, how many pounds of solids is in 1 week of backwash water?

Ans. 62,550 lb

9. The muffle furnace is set at 550°C. What is that temperature in Fahrenheit?

Ans. 1030°F

10. If a lake water with a total dissolved solids concentration of 150 mg/l (4 cfs) is supplemented with a well water having a total dissolved solids concentration of 800 mg/l (1 MGD), what is the total dissolved solids concentration of the mixed flow?

Ans. 332 mg/l

Nitrogen

1. Directions to prepare an ammonia stock solution specify dissolving 11.457 g of dried NH_4Cl in 1 l of distilled water. Then 10 ml of this stock solution is diluted to 1 l to make the standard.
   a. What is the concentration of nitrogen in the stock solution?
   b. What is the concentration of nitrogen in the standard?
   c. What is the concentration of ammonia in the stock solution?
   d. What is the concentration of ammonia in the standard?

Ans. a. 3025 mg/l
b. 30.25 mg/l
c. 3895 mg/l
d. 38.95 mg/l

2. In a colorimetric test for ammonia nitrogen, the sample absorbance was 0.18. The standard slope calculated out to 2.4, and the sample has been diluted four times with distilled water. What is the ammonia concentration in the original sample?

Ans. 1.73 mg/l

3. To prepare 250 ml of a potassium nitrate solution that is 100 mg/l nitrogen, how much of the chemical should be weighed out?

Ans. 180.4 mg

4. How many ml of a 0.2% solution of potassium nitrate are needed to prepare 100 ml of a 50-mg/l solution of potassium nitrate?

Ans. 2.5 ml

5. What is the mg/l concentration of a 1.3 *N* solution of ammonium hydroxide?

   Ans. 45,500 mg/l

6. Two samples are tested for nitrate. The following data is obtained:

Standards (mg/l concentration)	0.5	0.7	0.9
Standards (absorbance reading)	0.72	1.2	1.65
Sample 1 (absorbance)	2.84		
Sample 2 (absorbance)	0.98		

   Part of this test should be done over. What is wrong here? What should be done to correct it?

   Ans. Sample 1 absorbance is out of range of the standards. The sample should be diluted to read within the standards.

7. One liter of ammonium chloride solution has been prepared. Its concentration is 250 mg/l NH_3-N. How many mg of NH_3-N has been dissolved in the liter of water?

   Ans. 947 mg

8. A colorimetric procedure reads 500 micrograms of ammonia in the 50-ml sample. What is the ammonia concentration in mg/l?

   Ans. 10 mg/l

9. A stock nitrite solution is prepared at 100 mg/l. How many ml of this is needed to prepare 100 ml of a working standard at 100 μg/l?

   Ans. 0.1 ml

10. The potable water treatment plant is drawing from a river with a natural flow of 30 cfs. The normal nitrate content of the river water is 2 mg/l. Upstream, the 15 MGD wastewater treatment plant is discharging an effluent with a nitrate concentration of 25 mg/l. What is the nitrate concentration of the water that the low lift pumps are bringing into the potable water treatment plant?

    Ans. 12 mg/l

Phosphorus

1. From a 500-mg/l stock phosphate solution, how many ml would be needed to prepare 100 ml of a working standard of 2.5 mg/l?

   Ans. 0.5 ml

2. A colorimetric procedure for phosphorus is graphed and the standard slope is 264. Absorbance of the sample reads 0.023. What is the concentration of phosphorus in the sample?

   Ans. 6 mg/l

3. To prepare 1 l of a solution containing 1000 mg/l phosphate for a standard, dry $NaH_2PO_4{\cdot}5H_2O$ is used. How much of this chemical is needed?

   Ans. 2.212 g

4. An industrial phosphoric acid, as purchased, has a specific gravity of 2.8 and is 74% acid. How many ml would be needed to prepare 1 gal of a 2 *N* solution of it?

   Ans. 31.5 ml

5. In an analysis for phosphorus, the instrument reads 240 μg in this 20-ml sample. What is the concentration in mg/l?

   Ans. 12 mg/l

6. A 30 mg/l phosphate solution fills a 2-l flask. The entire contents of it is to be diluted to 2 mg/l. How many liters of the 2 mg/l solution will it make?

   Ans. 30 l

7. A solution of 1000 mg/l Na_2HPO_4 is what concentration of sodium phosphate, as phosphorus?

   Ans. 220 mg/l

8. Total phosphorus concentration of the process water of an 8 MGD treatment plant is 5.3 mg/l. Of this, 82% is orthophosphate. How many lb/day combined organic and polyphosphate are being processed?

   Ans. 63.7 lb/day

9. In preparing for a colorimetric procedure for orthophosphate, a purchased sodium phosphate standard of 1000 mg/l is to be used. Dilution of this to working standard concentrations of 1, 2, and 3 mg/l must be done. How?

   Ans. The purchased standard is too concentrated to dilute directly to working standard concentrations. Dilute it first to an intermediate concentration of 100 mg/l or 50 mg/l, and then down to 1, 2, and 3 mg/l.

10. A 20 MGD potable water treatment plant adds phospates for corrosion control at a concentration of 2 mg/l. How many lb/day are they adding to the flow of the city wastewater treatment plant?

    Ans. 333.6 lb/day

Atomic Absorption/Metals

1. Anhydrous ferric chloride was dissolved in a 100-ml flask to prepare an iron standard which is 6 mg/l iron. How many mg ferric chloride was needed?

   Ans. 1.7 mg

2. Testing for zinc, the sample is boiled down 50 times. Standards and samples are then read by atomic absorption. The sample absorbance is 0.02 and the standard slope is 100. What is the concentration of zinc in the original sample?

Ans. 0.04 mg/l

3. When diluting a sample for metals testing by 10, take 10-ml sample and bring up to 100 ml with distilled water. When diluting a sample by 8, take how many ml of it and bring up to 100 ml with distilled water?

Ans. 12.5 ml

4. A sodium sample tested colorimetrically was spiked with 2 mg/l sodium. If the unspiked sample had an absorbance of 0.23, the spiked sample had an absorbance of 1.15, and the standard slope was 2.2. What is the percent recovery of standard?

Ans. 99.5%

5. If 2 g of ferric chloride are dissolved into 1 gal of water, what is the mg/l concentration of iron in this solution?

Ans. 183 mg/l

6. A 500 mg/l stock solution of NaCl, as Cl has previously been prepared. To use that same solution to perform a test for sodium, what is the mg/l concentration of this NaCl solution, as Na?

Ans. 333 mg/l

7. How many ml of a solution composed of 1.2 g of sodium sulfate dissolved in 1 l of water are needed to prepare 1 l of a 10 mg/l solution of sodium sulfate, as Na?

Ans. 26 ml

8. Calcium was tested for by atomic absorption and yielded results of 45 mg/l, as calcium. What is this concentration registered as calcium carbonate?

Ans. 112.5 mg/l

9. What weight of dibasic potassium phosphate is needed to prepare 1 l of a 1000 mg/l solution of potassium?

Ans. 2.22 g

Appendix II

Preservation of Samples

Test	Preservation	Max. Time
Acidity	Refrigerate	24 h
Alkalinity	Refrigerate	24 h
Chloride	None required	No max.
Chlorine Residual	Analyze immediately	15 min
Color	Refrigerate	48 h
Specific Conductance	Refrigerate	28 days
Fluoride	None required	28 days
Hardness	Add HNO_3 to pH2	6 months
Metals	Add HNO_3 to pH2	No max.
Ammonia	Add H_2SO_4 to pH2	7 days
Nitrate	Refrigerate	48 h
Nitrite	Refrigerate	Do immediately
Organic Nitrogen	Refrigerate, add H_2SO_4 to pH2	7 days
Odor	Refrigerate	6 h
Dissolved Oxygen	Analyze immediately	8 h
pH	Analyze immediately	15 min
Phosphate	Refrigerate	48 h
Total Phosphorus	Add H_2SO_4 to pH2, refrigerate	28 days
Solids	Refrigerate	7 days
Sulfate	Refrigerate	28 days
Temperature	Analyze immediately	15 min
Turbidity	Refrigerate	24 h

Appendix III

Atomic Weights of Some Elements

Element	Atomic Weight	Element	Atomic Weight
Aluminum	27	Manganese	55
Calcium	40	Nitrogen	14
Carbon	12	Oxygen	16
Chlorine	35.5	Phosphorus	31
Copper	63	Potassium	39
Fluorine	19	Silicon	28
Hydrogen	1	Silver	108
Iodine	127	Sodium	23
Iron	56	Sulfur	32
Magnesium	24	Zinc	65

Appendix IV

Some Common Ions

Positive Ions	Negative Ions
H^+	Cl^-
Na^+	F^-
K^+	Br^-
NH_4^+	OCl^-
Ag^+	NO_2^-
Ca^{2+}	NO_3^-
Mg^{2+}	HCO_3^-
Ba^{2+}	OH^-
Cu^{2+}	CO_3^{2-}
Zn^{2+}	SO_4^{2-}
Hg^{2+}	O^{2-}
Fe^{2+} and $^{3+}$	S^{2-}
Mn^{2+} and $^{4+}$	PO_4^{3-}
Al^{3+}	
Si^{4+}	

Appendix V

Useful Conversions

Weight:	454 g	=	1 lb
	1000 μg	=	1 mg
	1000 mg	=	1 g
	1000 g	=	1 kg
Volume:	1000 ml	=	1 l
	3785 ml	=	1 gal
	7.48 ft^3	=	1 gal
Length:	1000 nm	=	1 μm
	1000 μm	=	1 mm
	1000 mm	=	1 m
Temperature:	20°C	=	68°F
	0°C	=	32°F
	100°C	=	212°F
For Water:	8.34 lb	=	1 gal
	1 ml	=	1 g
	1,000,000 mg	=	1 l

Bibliography

Adams, V., *Water and Wastewater Examination Manual.* Lewis Publishers, Chelsea, MI, 1990.

American Water Works Association, *Basic Science Concepts and Applications*, 2nd ed., Denver, CO, 1995.

American Public Health Association, American Water Works Association, and Water Environment Federation, *Standard Methods for the Examination of Water and Wastewater*, 20th ed., Washington, D.C., 1998.

American Public Health Association, American Water Works Association, and Water Pollution Control Federation, *Selected Chemical and Physical Standard Methods for Students*, Washington, D.C., 1986.

DeZuane, J., *Handbook of Drinking Water Quality*, 2nd ed., John Wiley & Sons, New York, 1997.

EPA, *Operation of Water Treatment Plants*, Vols. I and II. California State University, Sacramento, CA, 1983.

Faust, S. and Aly, O., *Chemistry of Water Treatment*, Butterworth, Woburn, MA, 1983.

Fountain, R., *Water Treatment Plant Operation: Chemistry for Operators*, Ann Arbor Science, Ann Arbor, MI, 1981.

Hach Co., *Water Analysis Handbook*, 2nd ed., Hach Co., Loveland, CO, 1992.

Hajian, Sr., H. and Pecsok, R., *Modern Chemical Technology*, Vol. I, Prentice-Hall, Englewood Cliffs, NJ, 1988.

Kenkel, J., *Analytical Chemistry for Technicians*, 2nd ed., Lewis Publishers, Boca Raton, FL, 1994.

Jackson, G., *Water Chemistry Manual for Water and Spentwater Personnel*, The Chemists Group, Inc., Oklahoma City, OK, 1990.

Manahan, S., *Environmental Chemistry*, 4th ed., Lewis Publishers, Chelsea, MI, 1990.

Masterton, W., Slowinski, E., and Stanitski, C., *Chemical Principles*, Saunders College Publishing, Philadelphia, 1983.

Sawyer, C. and McCarty, P., *Chemistry for Environmental Engineering*, McGraw-Hill, New York, 1978.

Index

A

B

C

Q

R

S

T

V

W

Z

For Product Safety Concerns and Information please contact our EU
representative GPSR@taylorandfrancis.com
Taylor & Francis Verlag GmbH, Kaufingerstraße 24, 80331 München, Germany

www.ingramcontent.com/pod-product-compliance
Lightning Source LLC
LaVergne TN
LVHW011130300926
849160LV00008B/590

* 9 7 8 1 5 6 6 7 0 4 8 6 1 *